Obesity and Esophageal Disorders

Obesity and Esophageal Disorders

Edited by

DHYANESH PATEL
Center for Swallowing and Esophageal Disorders, Vanderbilt University Medical Center, Nashville, TN, United States

ROBERT KAVITT
Center for Esophageal Diseases, The University of Chicago Medicine, Chicago, IL, United States

SHABNAM SARKER
Center for Swallowing and Esophageal Disorders, Vanderbilt University Medical Center, Nashville, TN, United States

MICHAEL VAEZI
Center for Swallowing and Esophageal Disorders, Vanderbilt University Medical Center, Nashville, TN, United States

ELSEVIER

ACADEMIC PRESS
An imprint of Elsevier

Academic Press is an imprint of Elsevier
125 London Wall, London EC2Y 5AS, United Kingdom
525 B Street, Suite 1650, San Diego, CA 92101, United States
50 Hampshire Street, 5th Floor, Cambridge, MA 02139, United States
The Boulevard, Langford Lane, Kidlington, Oxford OX5 1GB, United Kingdom

ISBN: 978-0-323-98365-5

Publisher: Stacy Masucci
Editorial Project Manager: Pat Gonzalez
Production Project Manager: Selvaraj Raviraj
Cover Designer: Matthew Limbert

Typeset by MPS Limited, Chennai, India

Contents

List of contributors

Joseph M. Blankush
Division of General Surgery, Vanderbilt University Medical Center, Nashville, TN, United States

Joseph R. Broucek
Division of General Surgery, Vanderbilt University Medical Center, Nashville, TN, United States

Pichamol Jirapinyo
Division of Gastroenterology, Hepatology and Endoscopy, Brigham and Women's Hospital, Boston, MA, United States; Harvard Medical School, Boston, MA, United States

Allon Kahn
Division of Gastroenterology and Hepatology, Mayo Clinic College of Medicine & Science, Scottsdale, AZ, United States

Rekha B. Kumar
Division of Endocrinology, Diabetes & Metabolism, Weill Cornell Medical College, New York, NY, United States

Joshua Lee
Division of Gastroenterology, Duke University School of Medicine, Durham, NC, United States

Benjamin Lloyd
Division of Gastroenterology, Duke University School of Medicine, Durham, NC, United States

Christopher P. Menzel
Division of General Surgery, Vanderbilt University Medical Center, Nashville, TN, United States

Rishi D. Naik
Division of Gastroenterology, Vanderbilt University Medical Center, Nashville, TN, United States

Adesola Oje
Division of Gastroenterology, Vanderbilt University Medical Center, Nashville, TN, United States

Nasim Parsa
Division of Gastroenterology and Hepatology, Mayo Clinic College of Medicine & Science, Scottsdale, AZ, United States

Amit Patel
Division of Gastroenterology, Duke University School of Medicine and the Durham Veterans Affairs Medical Center, Durham, NC, United States

Tiffany Patton
Department of Pediatrics, Section of Pediatric Gastroenterology, Hepatology, and Nutrition, University of Chicago, Chicago, IL, United States

Francesca Raffa
Division of Gastroenterology, Vanderbilt University Medical Center, Nashville, TN, United States

Shakirat Salvador
Division of Gastroenterology, Vanderbilt University Medical Center, Nashville, TN, United States

Shabnam Sarker
Division of Gastroenterology, Vanderbilt University Medical Center, Nashville, TN, United States

Akinari Sawada
Wingate Institute of Neurogastroenterology, Blizard Institute, Barts and The London School of Medicine and Dentistry, Queen Mary University of London, London, United Kingdom

Ilia Sergeev
Wingate Institute of Neurogastroenterology, Blizard Institute, Barts and The London School of Medicine and Dentistry, Queen Mary University of London, London, United Kingdom

Kevin Shah
Department of Internal Medicine, Vanderbilt University Medical Center, Nashville, TN, United States

Raj Shah
Division of Gastroenterology, Hepatology and Endoscopy, Brigham and Women's Hospital, Boston, MA, United States; Harvard Medical School, Boston, MA, United States

Daniel Sifrim
Wingate Institute of Neurogastroenterology, Blizard Institute, Barts and The London School of Medicine and Dentistry, Queen Mary University of London, London, United Kingdom

Okeefe L. Simmons
Division of Gastroenterology, Weill Cornell Medical College, New York, NY, United States; Division of Endocrinology, Diabetes & Metabolism, Weill Cornell Medical College, New York, NY, United States

Matthew D. Spann
Division of General Surgery, Vanderbilt University Medical Center, Nashville, TN, United States

Gitanjali Srivastava
Division of Diabetes, Endocrinology & Metabolism, Department of Medicine, Vanderbilt Weight Loss Center, Vanderbilt University School of Medicine, Nashville, TN, United States; Department of Surgery, Vanderbilt University School of Medicine, Nashville, TN, United States; Department of Pediatrics, Vanderbilt University School of Medicine, Nashville, TN, United States

Christopher C. Thompson
Division of Gastroenterology, Hepatology and Endoscopy, Brigham and Women's Hospital, Boston, MA, United States; Harvard Medical School, Boston, MA, United States

Joseph Wawrzynski
Department of Medicine, Duke University School of Medicine, Durham, NC, United States

Preface

It is our great pleasure to introduce you to an issue of *Obesity and Esophageal Disorders.*

Obesity is commonly encountered in our medical practice and is linked to numerous comorbidities along with severe impairment of overall health. According to the National Institute of Health, two of three patients are either overweight or obese. The WHO has found that over 650 million patients are affected by obesity and nearly 2 billion who are overweight. Thus obesity is a growing medical and public health problem worldwide with a staggering $147 billion in medical costs. Overweight and obesity are well-known risk factors for a variety of gastrointestinal disorders but have directly contributed to an exponentially rising incidence of gastroesophageal reflux disease (GERD) with a recent study showing that reflux symptoms affect one in three Americans. This has also affected the incidence of long-term complications of GERD including esophageal dysmotility, Barrett's esophagus, and esophageal cancer. There is a critical gap in awareness about impact of obesity on esophageal disorders and medical provider's ability to provide timely diagnosis and management.

The field of esophageal diseases has been transformed over the last decade with the advent of newer diagnostic technologies (high-resolution manometry and functional luminal imaging probe for motility disorders; wireless pH monitoring; and mucosal integrity devices for GERD) and treatment options (bariatric surgery, endobariatrics, and medical) that decrease the latency period for diagnosis and management of these patients. Furthermore, as surgical management of obesity is increasing in prevalence, we are also now starting to understand the potential effects of different bariatric surgeries on esophageal motility.

This book is the first comprehensive, state-of-the art review on impact of obesity on various esophageal disorders and how to approach, recognize, and treat those disorders. This book will serve as a valuable resource for clinicians, surgeons, researchers, and trainees with an interest in the management of patients with obesity. The book provides an exhaustive literature review of all the current evidence behind diagnosis, evaluation, and management of various obesity-related esophageal disorders written by the most prominent clinicians and researchers in this field.

The chapters are organized so that the readers systematically learn about prevalence of obesity and impact on gastrointestinal health in both adults and pediatric population followed by impact on GERD and Barrett's esophagus. We subsequently review how obesity and bariatric surgery can affect esophageal motility and discuss diagnostic tools. Lastly, we include nonsurgical, endoscopic, and surgical therapeutic options in patients with obesity and esophageal disorders. We are grateful to the contributors and hope that the book provides insight into an evidence-based approach to taking care of this patient population.

Dhyanesh A. Patel
Robert Kavitt
Shabnam Sarker
Michael Vaezi

CHAPTER 1

Obesity and its impact on gastrointestinal health

Shakirat Salvador, Adesola Oje and Shabnam Sarker
Division of Gastroenterology, Vanderbilt University Medical Center, Nashville, TN, United States

Overview

Obesity is a complex disease that can predispose to metabolic derangements that can lead to multiorgan dysfunction. This chapter will review the epidemiology of obesity, describe the economic burden of disease, and review the complex association between metabolic syndrome and its role in gastrointestinal diseases. Finally, the impact of obesity on the pediatric population will be discussed.

Introduction

Obesity is defined as a body mass index (BMI) of greater than or equal to 30, with further subgroup classifications defined by the World Health Organization (WHO). Class I obesity is defined as a BMI between 30 and 34.9. Class II obesity is defined as BMI between 35 through 39.9, and Class III obesity is defined as BMI greater than or equal to 40.[1] Obesity leads to metabolic neurohormonal disarray that leads to alterations in lipid metabolism and lipid resistance. The metabolic syndrome is characterized by increased waist circumference, dyslipidemia, hypertension, and hyperglycemia.[2] This is exceedingly recognized in other metabolic conditions such as diabetes, cardiovascular disease, cerebrovascular disease, and malignancies. Other disease associations include obstructive sleep apnea and polycystic ovarian syndrome. In essence, metabolic syndrome describes a complex interplay of metabolic derangements with established comorbid conditions.[2]

The effects of obesity and its relationship with gastrointestinal and hepatologic conditions continue to evolve. This becomes particularly evident in conditions such as gastroesophageal reflux disease, dyspepsia, nonalcoholic

Obesity and Esophageal Disorders
DOI: https://doi.org/10.1016/B978-0-323-98365-5.00006-1

fatty liver disease, small intestinal bacterial overgrowth, diverticulosis, and pancreatitis.[2]

Epidemiology

The international prevalence of obesity continues to rise. According to the WHO, the current worldwide prevalence of obesity is almost triple that of 1975. In 2016, approximately 1.9 billion (39%) adults aged 18 and older were overweight, while 650 million (19%) of this cohort were obese. In this population, 39% were men while 40% were women.[3] Currently, it is estimated that 39%–49% of the global population is either obese or overweight.[4] Across all countries, the prevalence of obesity among women has been on the rise. In the 1970s, data looking at the overall global trend of obesity showed a prevalence of greater than 5% but less than 20% of women in developed nations.[5]

According to the United States Center for Disease Control, the prevalence of obesity within the United States was 42.4% with a projected increase to up to 48.9% by the year 2030. The most recent data published in 2020 by the National Center for Health Statistics highlight this trend. Between 2017 and 2018, 42.4% of the adult population was noted to be obese. Data for various ethnic groups consistently demonstrate that Non-Hispanic Black Americans had the highest prevalence of obesity 49.6% when compared to non-Hispanic Asians 17.5%, non-Hispanic Caucasians 42.2%, and 44.8% in Hispanics.[6] These numbers are largely impacted by racial disparities, stigmatization of weight, and psychosocial stressors, as well as factors that exacerbate socioeconomic inequalities.[7]

Fig. 1.1 shows an illustrated representation of the percent of adults in the United States who have obesity in the year 2020 based on data compiled by the Center for Disease Control using the Behavioral Risk Factor Surveillance System.[8]

Medical Complications

As the burden of disease increases, there have been multiple efforts to understand the effects of obesity and its contributions to metabolic alterations and organ dysfunction. Excess weight and obesity are known risk factors for several chronic medical conditions including cardiovascular disease, diabetes mellitus, chronic kidney disease, malignancies, musculoskeletal diseases, stroke, dementia, nonalcoholic fatty liver disease, and

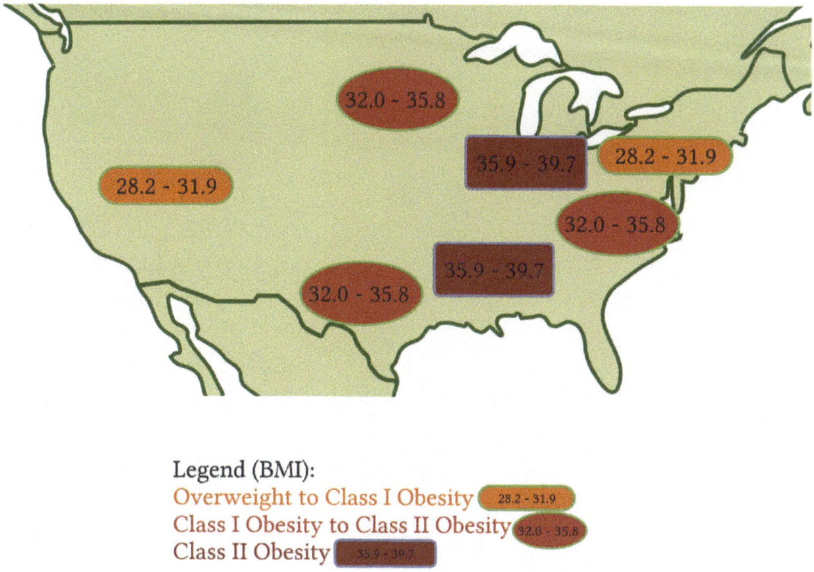

Figure 1.1 Percent of adults in the United States who have obesity in the year 2020 based on data compiled by the Center for Disease Control using the Behavioral Risk Factor Surveillance System.[8] *Created wih BioRender.com.*

obstructive sleep apnea.[9] Fig. 1.2 shows an illustrated representation of the organs and associated disorders linked to obesity.

Several studies have highlighted the impact of obesity on the cardiovascular system. The Global Burden of Disease in 2015 showed that people with high BMI accounted for 4.0 million deaths in 2015 with greater than 2/3 secondary to cardiovascular disease even after accounting for critical illness and tobacco use.[1,10] Globally, cardiovascular disease is the leading cause of death in the overweight and obese population accounting for 2.7 million deaths (95% CI: 1.8−3.7) and 66.3 million disability-adjusted life years (95% CI: 45.3−88.5). As a result, there are currently several studies assessing the impact of weight loss and lifestyle modification on cardiovascular outcomes. The Look Action for Health in Diabetes (AHEAD) was a multicenter randomized controlled trial conducted between 2001 and 2012 to determine the impact weight loss had on cardiovascular morbidity and mortality in patients with T2DM (the Look Ahead Research Group).[11] Although results failed to show a reduction in Major Adverse Cardiovascular Events (MACE) or cardiovascular mortality after 9.61 years, post-hoc analysis showed that patients with $\geq 10\%$ weight loss had significant reductions in cardiac events.[11]

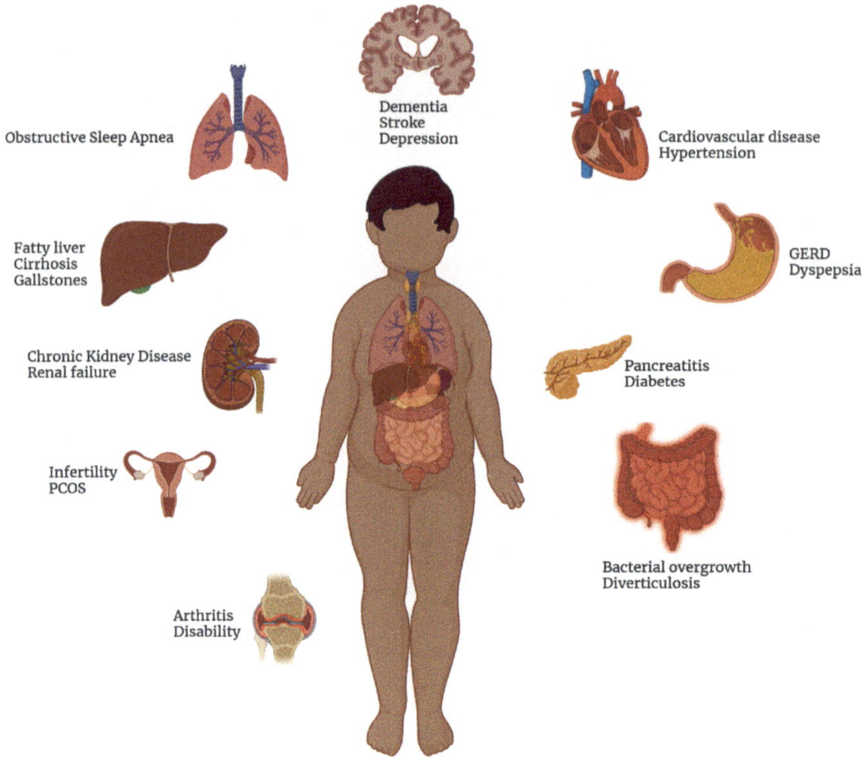

Figure 1.2 End organ damage due to obesity. *Created with BioRender.com.*

Physical exercise, particularly aerobic exercise, and pharmacologic interventions such as the glucagon-like peptide 1 agonists have been shown to reduce MACE.[12]

Obesity and Pharmacokinetics

Obesity also impacts drug pharmacokinetics, especially those of lipophilic medications.[13] While oral absorption is not affected, the distribution and clearance of certain drugs are affected. Each medication presents a unique challenge as its affinity for adipose tissue is unique.[13] This requires closer monitoring of such medication in obese patients and can present challenges in reaching appropriate dosing.

Economic Burden of Obesity

Obesity poses an economic burden not only to individuals and families but nations as well. The costs incurred by obesity are either direct, those associated with medical services (e.g., the value of lost work, insurance costs, wages), or indirect, costs due to resources forgone.[14]

With the rising prevalence of obesity, the economic costs have continued to rise. It is expected that obesity-related costs will rise by $48–66 billion/year by 2030.[14]

Esophagogastric Disease

Introduction

Gastroesophageal reflux disease (GERD) is a condition that describes a constellation of symptoms that are a result of physiologic and anatomical deficiencies that lead to the abnormal retrograde flow of gastric contents to the esophagus. Over time, this contact can lead to esophageal injury that can result in esophageal dysmotility, esophagitis, Barrett's esophagus, esophageal strictures, esophageal ulcerations, and esophageal malignancy.[15] Typical symptoms include heartburn, regurgitation, and dysphagia. Worldwide epidemiologic studies consistently demonstrate the association between GERD and obesity.[2] Extraintestinal manifestations of GERD also include chronic cough, noncardiac chest pain, hoarseness, asthma, reflux laryngitis, and odontologic decay.

Epidemiology

The estimated worldwide prevalence of GERD in adults is approximately 13.3%. The highest rates of GERD are within South Asia, Central America, South America, Europe, and North America. The prevalence of GERD is approximately 73% in patients with obesity.[15] The economic cost burden is also substantial, with an estimated cost of over $12.3 billion within the United States for GI prescription drugs, $7.7 billion for GERD alone.[16]

Pathophysiology

GERD occurs when there is dysfunction between the gastroesophageal junction, the lower esophageal sphincter (LES), and the crural diaphragm. Specifically, the length and resting pressure of the LES are integral components of the prevention of reflux. This is what makes up the so-called "antireflux barrier." Anatomic aberrations such as a hiatal hernia are also key risk factors for the development of GERD. Alterations in intraabdominal pressure can influence the function of the antireflux barrier.

Truncal adiposity can increase intraabdominal pressure that contributes to increased pressure gradients that negatively impact the antireflux barrier. In addition, truncal adiposity also increases the risk of the development of hiatal hernias.[15]

Manifestations of Disease

Typical symptoms of GERD include heartburn and regurgitation. Heartburn often occurs in the postprandial setting and is classically described as a substernal burning sensation. This burning sensation can also radiate across the chest, toward the back, and proximally to the neck and mouth. Symptoms can be exacerbated by positioning and can precipitate nocturnal symptoms. Regurgitation is described as the sensation of the retrograde flow of gastric contents. With prolonged GERD, patients can develop complications of the disease including erosive esophagitis, esophageal strictures, Barrett's esophagus, and esophageal adenocarcinoma.[15] Patients who are obese have 2.4 times increased risk of developing esophageal cancer, and risk rates increased concordantly by 16% per 1 kg/m^2 increase in BMI.[2,17]

Dyspepsia describes a constellation of postprandial symptoms. Characteristics of dyspepsia epigastric abdominal symptoms include discomfort, pain, heartburn, early satiety, and postprandial fullness. Other commonly associated symptoms described include eructation, nausea, and/or vomiting. Dyspepsia can be a manifestation of GERD in up to 20% of cases. Dyspepsia is also associated with functional disease, peptic ulcer disease, upper GI malignancy, pancreaticobiliary disease, and lower GI tract disease.

Hepatology

Introduction

Nonalcoholic fatty liver disease (NAFLD) occurs when there is an excess storage of fat within the liver. This process, called steatosis, is pathologic when there it affects more than 5% of hepatocytes in the absence of alternative etiologies.[18] Similar to the rising prevalence of obesity, the incidence of NAFLD continues to rise.[19] Currently, NAFLD affects a quarter of the world's population.[18] The complications of NAFLD include nonalcoholic steatohepatitis (NASH), end-stage liver disease (cirrhosis), and hepatocellular carcinoma (HCC). In addition to the morbidity associated with the manifestations of these hepatic complications, there is increased mortality of cardiovascular disease and malignancy within this group.[19]

Epidemiology

There is a complex causal link between that of the metabolic syndrome and the development of NAFLD.[19,20] The prevalence of NAFLD within the Unites States continues to increase. One of the complications of NAFLD is HCC. There is a rising prevalence of HCC, with a rising annual incidence of 9%. Given the rising prevalence and international burden of disease, therapeutic interventions are focused on dietary modifications, lifestyle interventions, medical therapy, surgical therapy, and optimization of medical comorbidities.[18]

Pathophysiology

NAFLD describes a spectrum of diseases. When steatosis accumulates within hepatocytes, inflammation can cause progressive liver imaging resulting in NASH. There is a complex interrelationship that influences inflammation that is mediated by proinflammatory cytokine production. Over time, continued injury can promote ongoing inflammation, hepatocyte necrosis, and fibrosis. However, the spectrum of disease can vary from stability, regression, or progression.[18] Repeated injury can result in cirrhosis. The diagnosis of NAFLD can be made noninvasively by clinical and radiographic evaluation. Additionally, liver biopsy remains the gold standard in diagnosis.[18,20]

Diagnosis

The earliest signs of the disease include abnormal aminotransferases (AST, ALT) and/or abnormal imaging. Therefore diagnosis of NAFLD can be achieved both with a combination of biochemical analysis and radiographic evaluation. Radiographic diagnosis can be achieved via abdominal ultrasound (US), computed tomography (CT), or magnetic resonance imaging. Findings demonstrate hepatic steatosis. There are noninvasive scoring tools that are available to assess risk factors for advanced disease. Some commonly used scoring tools incorporate age, BMI, blood glucose, aminotransferase ratios, platelet counts, and albumin. Noninvasive measurements of steatosis and fibrosis can also be achieved via vibration-controlled transient elastography, acoustic radiation force impulse, or magnetic resonance spectroscopy. The gold standard in the identification of hepatic parenchymal disease, quantification of steatosis, and staging of fibrosis remains the liver biopsy. The liver biopsy core specimen can be obtained via a percutaneous liver biopsy (CT guided, US-guided, or nonimaging guided), transjugular liver biopsy, or endoscopic ultrasound-guided liver biopsy.[20]

Manifestations of Disease

Given that NAFLD encompasses a spectrum of diseases, the manifestations of the disease are also broad. Patients remain asymptomatic until their disease advances to end-stage liver disease. Therefore patients can often present with manifestations of decompensated liver disease including complications of portal hypertension (ascites, spontaneous bacterial peritonitis, volume overload, hepatic hydrothorax, variceal hemorrhage), hepatic encephalopathy, portopulmonary hypertension, hepatopulmonary syndrome, and hepatorenal syndrome. HCC is a complication of NAFLD, with NAFLD being the most common risk factor for the development of HCC. However, most cases of HCC occur in the setting of cirrhosis.[20]

Obesity and the Gut Microbiome

The role of the gut microbiome is a rapidly evolving field. The effects of the gut microbiome across the pathophysiology of various disease states are still being evaluated. Specifically, the contributions of environmental mediators are introduced via diet and are therefore dependent on cultural preferences and socioeconomic status.[21] With regards to metabolic disease, it is postulated that disruption of the gut microbiome contributes to metabolic changes and pro-inflammatory states, particularly on lipid metabolism.[21] This has been demonstrated in mouse studies. Germ-free mice, as compared to conventional mice had lower body weights. In addition, conventional mice have increased rates of weight gain. Fecal microbiota transplants (FMT) were utilized to study the role of the gut microbiome and its effects on obesity. In mice who had FMT from obese human donors induced higher amounts of weight gain. Therefore the gut microbiome provides a modulatory effect on lipid metabolism and adipose storage.[21] Further studies will be helpful to determine the role of FMT in weight gain in those with and without infectious *Clostridium difficile* as the indication for FMT.

Functional Gastrointestinal Disorders

Obesity has been correlated with functional GI disorders and bowel changes. There are strong links with diarrhea thought to be related to ingestion of high fat and difficult to absorb sugars leading to osmotic diarrhea.[22] Diets heavy in high fructose corn syrup, for example, can cause such GI upset.[23] Increased GERD in obesity inevitably leads to increased PPI use. PPIs have been associated with small intestinal bacterial overgrowth that in turn can drive irritable bowel syndrome.[23] Additionally, diets low in fiber and high in processed food, can predispose to constipation. Binge eating can also lead to abdominal pain and IBS.[23]

Pediatric Population

Once thought to be a disease that plagued high-income countries, data currently show alarming numbers in developing nations. Compared to the year 2000, the incidence of obesity has increased by 24% in children less than age 5 in Africa. At the same time, about half of children in the same age group are either obese or overweight in Asia.[3] The Global Study of Disease (GSD), a consortium that provides information regarding health loss from hundreds of diseases and injuries has repeatedly identified high BMI as one of the leading (Top 5) risk factors for deaths and disability-adjusted life years.

Esophageal physiology

Eating can be broken down into four main phases: oral, triggering the swallowing reflex, pharyngeal phase where the bolus is transported to the pharynx, and the esophageal phase—where the bolus is transferred to the stomach. In the neonatal and infant stage, all stages are involuntary and reflexive. With time, the oral phase becomes voluntary with some control of the swallow reflex in later life.[24]

Conclusion

Obesity is a very complex and rapidly growing chronic medical condition leading to increased mortality. Obesity and its role in GI disorders will be explored in this book. With greater knowledge and understanding, we hope to be able to better care for our expanding obesity epidemic.

Disclosure

The authors of this chapter have no financial disclosures to report.

References

1. Global BMIMC, Di Angelantonio E, Bhupathiraju SN, et al. Body-mass index and all-cause mortality: individual-participant-data *meta*-analysis of 239 prospective studies in four continents. *Lancet*. 2016;388(10046):776–786.
2. Nam SY. Obesity-related digestive diseases and their pathophysiology. *Gut Liver*. 2017;11(3):323–334.
3. WHO. Obesity and overweight. <https://www.who.int/news-room/fact-sheets/detail/obesity-and-overweight\>; 2021. Accessed.
4. Maffetone PB, Rivera-Dominguez I, Laursen PB. Overfat and underfat: new terms and definitions long overdue. *Front Public Health*. 2016;4:279.
5. Jaacks LM, Vandevijvere S, Pan A, et al. The obesity transition: stages of the global epidemic. *Lancet Diabetes Endocrinol*. 2019;7(3):231–240.
6. CDC. Overweight & Obesity. <https://www.cdc.gov/obesity/data/adult.html\>; 2021. Accessed.
7. Bell CN, Kerr J, Young JL. Associations between obesity, obesogenic environments, and structural racism vary by county-level racial composition. *Int J Env Res Public Health*. 2019;16(5).
8. CDC. Nutrition, physical activity, and obesity: data, trends and maps. <https://www.cdc.gov/nccdphp/dnpao/data-trends-maps/index.html\>; 2021. Accessed.
9. Field AE, Coakley EH, Must A, et al. Impact of overweight on the risk of developing common chronic diseases during a 10-year period. *Arch Intern Med*. 2001;161(13):1581–1586.
10. Afshin A, Forouzanfar MH, Reitsma MB, et al. Health effects of overweight and obesity in 195 countries over 25 years. *N Engl J Med*. 2017;377(1):13–27.
11. Gregg EW, Jakicic JM, Blackburn G, et al. Association of the magnitude of weight loss and changes in physical fitness with long-term cardiovascular disease outcomes in overweight or obese people with type 2 diabetes: a post hoc analysis of the Look AHEAD randomised clinical trial. *Lancet Diabetes Endocrinol*. 2016;4(11):913–921.
12. Powell-Wiley TM, Poirier P, Burke LE, et al. Obesity and cardiovascular disease: a scientific statement from the American heart association. *Circulation*. 2021;143(21):e984–e1010.
13. Hanley MJ, Abernethy DR, Greenblatt DJ. Effect of obesity on the pharmacokinetics of drugs in humans. *Clin. Pharmacokinet*. 2010;49(2):71–87.
14. Wang YC, McPherson K, Marsh T, Gortmaker SL, Brown M. Health and economic burden of the projected obesity trends in the USA and the UK. *Lancet*. 2011;378(9793):815–825.
15. Maret-Ouda J, Markar SR, Lagergren J. Gastroesophageal reflux disease: a review. *Jama*. 2020;324(24):2536–2547.
16. Hanauer SB. The burdens of digestive diseases. *Nat Rev Gastroenterol Hepatol*. 2009;6(7):377.
17. Thrift AP, Shaheen NJ, Gammon MD, et al. Obesity and risk of esophageal adenocarcinoma and Barrett's esophagus: a Mendelian randomization study. *J Natl Cancer Inst*. 2014;106(11).
18. Younossi ZM, Marchesini G, Pinto-Cortez H, Petta S. Epidemiology of nonalcoholic fatty liver disease and nonalcoholic steatohepatitis: implications for liver transplantation. *Transplantation*. 2019;103(1):22–27.

19. Younossi ZM, Corey KE, Lim JK. AGA clinical practice update on lifestyle modification using diet and exercise to achieve weight loss in the management of nonalcoholic fatty liver disease: expert review. *Gastroenterology*. 2021;160(3):912–918.
20. Chalasani N, Younossi Z, Lavine JE, et al. The diagnosis and management of nonalcoholic fatty liver disease: practice guidance from the American association for the atudy of liver diseases. *Hepatology*. 2018;67(1):328–357.
21. Aron-Wisnewsky J, Warmbrunn MV, Nieuwdorp M, Clément K. Metabolism and metabolic disorders and the microbiome: the intestinal microbiota associated with obesity, lipid metabolism, and metabolic health-pathophysiology and therapeutic strategies. *Gastroenterology*. 2021;160(2):573–599.
22. Aro P, Ronkainen J, Talley NJ, Storskrubb T, Bolling-Sternevald E, Agréus L. Body mass index and chronic unexplained gastrointestinal symptoms: an adult endoscopic population based study. *Gut*. 2005;54(10):1377–1383.
23. Ho W, Spiegel BM. The relationship between obesity and functional gastrointestinal disorders: causation, association, or neither? *Gastroenterol Hepatol (NY)*. 2008;4(8):572–578.
24. Dodrill P, Gosa MM. Pediatric dysphagia: physiology, assessment, and management. *Ann Nutr Metab*. 2015;66(Suppl 5):24–31.

CHAPTER 2

Impact of obesity on esophageal physiology in pediatrics

Tiffany Patton
Department of Pediatrics, Section of Pediatric Gastroenterology, Hepatology, and Nutrition, University of Chicago, Chicago, IL, United States

Introduction

Childhood encompasses a period of rapid growth and development physically, emotionally, and socially from birth to young adulthood. During this period the body undergoes significant maturation of physiologic function. This is particularly evident in the gastrointestinal tract, which is one of most extensive, diverse, and complex organ systems within the human body. The gastrointestinal tract will ultimately multiply its physical length nearly threefold from birth to adulthood, while performing the physiologic functions of nutrient digestion, absorption, and elimination of waste. The magnitude of the physical capacity and function of the gastrointestinal tract also make it susceptible to numerous pathologies, many of which are common in pediatrics and adults. Gastrointestinal symptoms may also develop in response to several extraintestinal conditions and diseases, such as obesity, autoimmune disorders, and neurological impairment. Specifically, obesity has numerous effects on the gastrointestinal tract and has a strong association with adult gastroesophageal reflux (GER) disease (GERD). The prevalence of obesity in adults in the United States is ~42%, while the prevalence of adult GERD in the United States is ~20%.[1–3] Adult data shows that obese people are nearly twice as likely to suffer from GERD symptoms compared to nonobese people.[4] However, data establishing a correlation between obesity and GERD in children is less abundant and conflicting.

The goals of the present chapter are to describe the prevalence of GERD throughout childhood as well as to discuss the most recent recommendations for management of pediatric GERD. The chapter will also identify current trends in pediatric obesity prevalence and discuss the pathophysiology of obesity-induced GERD. Lastly, the chapter will

Obesity and Esophageal Disorders
DOI: https://doi.org/10.1016/B978-0-323-98365-5.00011-5

discuss current recommendations of childhood obesity treatment and identify long-term complications of untreated obesity-induced GERD in children.

Physiologic function of the esophagus

Derived from the anterior portion of the embryonic foregut, the esophagus is composed of stratified squamous epithelial mucosa and a combination of striated and smooth muscle.[5] The esophagus will elongate from ~8 cm at birth to ~25 cm in adulthood and its major physiologic purpose to propel liquid and solid boluses from the oropharynx into the stomach is established in utero.[6] The act of swallowing requires synchronized contraction/relaxation of various esophageal zones and antegrade muscular peristalsis that is accomplished by the coordination of three esophageal regions: the upper esophageal sphincter, esophageal body, and the lower esophageal sphincter (LES). The upper esophageal sphincter is a high-pressure zone that serves as a proximal barrier to prevent gastrointestinal reflux into the laryngopharynx during esophageal peristalsis and to avoid excessive passage of air into the esophagus during inhalation. The esophageal body accounts for most of the esophageal length and extends from the cervical region, through the thorax, and terminates in the upper abdomen. The esophageal body is composed of skeletal muscle in the proximal region with a gradual transition to smooth muscle distally and remains relaxed at rest. Peristalsis is the sequential contraction of esophageal muscles moving distally at a childhood velocity of 3.0 cm/s along the esophageal body before reaching the LES. The esophageal body is capable of primary peristalsis, initiated by swallowing, and secondary peristalsis, elicited in response to luminal distension by fluid and air.[6] The LES is a high-pressure zone located at the gastroesophageal junction (GEJ), which serves to prevent gastric refluxate into the esophagus, but may also relax to allow retrograde passage of air or fluid during eructation or vomiting, respectively.

There are a variety of pediatric conditions and diseases that disrupt the natural physiology of the esophagus. Neuromuscular disorders, such as muscular dystrophies and achalasia, may alter esophageal motility and contribute to symptoms of deglutition and dysphagia.[7,8] Connective tissue disorders, such as Sjogren's syndrome and scleroderma, may present as dysphagia due to abnormal esophageal motility and fibrosis, as well as decreased LES pressure.[9] Inflammatory conditions, such as Crohn's disease, eosinophilic esophagitis, dermatomyositis, and reflux esophagitis, may contribute to esophageal edema exacerbating dysphagia, altered motility, and the development of strictures.[10] Additional conditions inclusive of neurological impairment, esophageal infections, and obesity have been associated with altered esophageal physiology and may predispose to increased occurrences of GERD.

Gastroesophageal reflux disease

GER is defined as the involuntary passage of gastric contents into the esophagus with or without regurgitation and/or vomiting.[11] Three major factors contribute to the occurrence of GER: (1) transient relaxation of the LES (TRLES), (2) anatomic disruption of the GEJ, and (3) and an incompetent LES.[12] These factors are most evident in preterm infants born less than 34-week gestation who may have a GER incidence of 22% due to predisposing risk factors of physiological immaturity of the LES, impaired esophageal peristalsis, high volume of milk intake, and slower gastric emptying.[13] GER is considered a normal phenomenon that occurs in children and adults, presenting in ~70% of infants younger than 7 months and typically resolving by 2 years of age in ~95% of children.[11,14,15] GERD occurs when GER is associated with bothersome symptoms and/or the development of esophageal complications (i.e., strictures, erosive esophagitis).[11]

In North America the prevalence of adult GERD has risen since 1995 and is now estimated to be as high as 27.8%, in comparison to 7% in East Asia, 33% in the Middle East, and 51.2% in Greece.[2,4] Publications with pediatric GERD prevalence data are limited; however, a recent systematic review reported a pooled prevalence of 26.9% for GERD symptoms in infants younger than 18 months.[16] This prevalence was noted to decrease from 25.5% in infants at 1 month of age to <3% at 6 months.[17] A Japanese study reported an overall prevalence of 3.2% for weekly GERD symptoms in children less than 10 years of age.[18] While the prevalence of weekly GERD symptoms in children aged 10–21 years is estimated to be 10.1% worldwide.[16] To date, there has not been an associated increased risk for GERD in children based on ethnicity and few studies have indicated a higher risk of GERD symptoms in females compared to males.[19–21] Several studies have shown an emerging association between obesity and GERD in children, similar to that seen in adults.[16,21–24] In addition, GERD is also more prevalent among children with neurological impairment, congenital heart disease, congenital diaphragmatic hernia, chromosomal abnormalities, asthma, cystic fibrosis, interstitial lung disease, and abnormalities of the gastrointestinal tract.[13,25]

Clinical symptoms of GERD may manifest differently depending on the age, neurological status, and maturation of the child. In infants, symptoms of recurrent vomiting, poor weight gain, back arching, feeding refusal, irritability/fussiness, sleep disturbances, respiratory symptoms, and dystonic neck posturing (Sandifer syndrome) may correlate with the presence of GERD. Older children with GERD, being capable of describing their symptoms, may present with heartburn, epigastric pain, dysphagia, recurrent vomiting and/or regurgitation, asthma, recurrent pneumonia, dental erosions, and various upper airway symptoms.[11,26] Specific "red flag" symptoms beyond those commonly associated with GERD, such as weight loss, fever, seizures, nocturnal vomiting, hematemesis, abdominal distension, diarrhea, rectal bleeding, or dysuria, warrant further evaluation for other organic illnesses. Prolonged esophageal exposure to gastric acid in untreated GERD can lead to more severe complications, such as erosive esophagitis, Barrett esophagus, and adenocarcinoma of the esophagus.[27]

In 2018 the North American Society for Pediatric Gastroenterology, Hepatology, and Nutrition (NASPGHAN) published updated clinical practice guidelines for pediatric GERD, which modernized the diagnostic and therapeutic management of GERD in children. These guidelines suggest that the diagnosis of pediatric GERD should be based upon a thorough clinical history of characteristic symptoms in the absence of "red flag" symptoms. In addition, the guidelines propose the empiric treatment of GERD symptoms with thickened formula or breast milk feeds in infants and a 4- to 8-week trial of empiric proton-pump inhibitor (PPI) therapy for children (see Fig. 2.1).[11] There are several diagnostic tests available to help elucidate GERD causality and evaluate "red flag" symptoms. Previously considered the gold-standard diagnostic test for pediatric GERD is the 24-hour esophageal pH monitoring study which measures the cumulative percentage of time within a 24-hour period that the esophageal pH is less than 4 (also known as the reflux index or RI). RIs > 12% in infants and >7% older children are considered abnormal, while RIs between 3% and 7% are intermediate, and RIs < 3% are normal.[14] Twenty-four-hour pH monitoring can be performed with multichannel intraluminal impedance measurement, which enhances the detection of acidic and nonacidic reflux events that can be correlated with

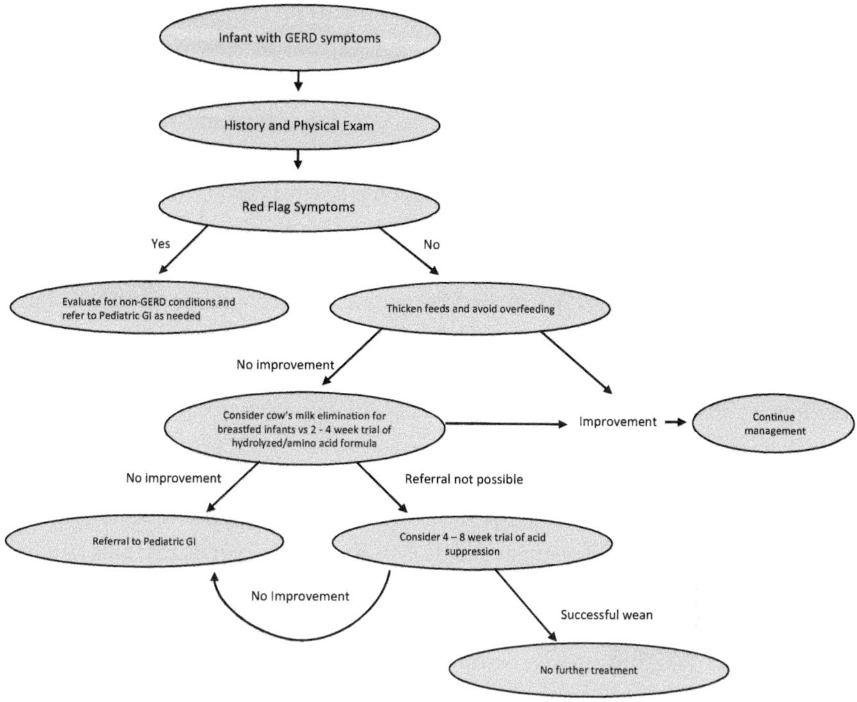

Figure 2.1 Infantile GERD management algorithm. *GERD*, gastroesophageal reflux disease. *Adapted from Rosen R, Vandenplas Y, Singendonk M, et al. Pediatric Gastroesophageal Reflux Clinical Practice Guidelines: Joint Recommendations of the North American Society for Pediatric Gastroenterology, Hepatology, and Nutrition and the European Society for Pediatric Gastroenterology, Hepatology, and Nutrition. J Pediatr Gastroenterol Nutr 2018;66(3):516–554. doi: 10.1097/MPG.0000000000001889. PMID: 29470322; PMCID: PMC5958910.*

GERD symptoms. Radiographic upper GI contrast studies are helpful to detect anatomical anomalies contributing to GERD, such as hiatal hernia, malrotation, microgastria, and antral webs, but do not confirm GERD, as physiologic GER is often witnessed during these studies. Upper endoscopy is not required to confirm GERD but may be utilized to evaluate for upper gastrointestinal mucosal inflammatory conditions, such as erosive esophagitis, eosinophilic esophagitis, and gastritis. Esophageal manometry is typically not indicated in the evaluation of GERD, unless there is a need to assess esophageal motility. Other diagnostic modalities, such as ultrasonography, nuclear scintigraphy, and salivary pepsin detection, are not currently recommended for the diagnosis of pediatric GERD.

Treatment of pediatric GERD may vary depending on the age of the patient but will typically consist of behavioral modification with antireflux precautions, dietary modification, antacid therapy, or surgical intervention. In infants with GERD, nonpharmacologic interventions such as the use of thickening agents for formula or breast milk, avoidance of overfeeding, and a 2- to 4-week trial of an extensively hydrolyzed protein or amino-acid formula can be attempted as first-line therapy. In older children with GERD, lifestyle changes inclusive of eating smaller meals and dietary avoidance of caffeine, chocolate, carbonated beverages, and spicy foods may offer symptomatic relief. If children are overweight or obese, weight loss may be recommended to alleviate GERD symptoms. Positional changes such as a head of bed elevation and left-lateral decubitus may also reduce GERD symptoms.[11]

The goal of pharmacological treatments utilized for pediatric GERD is to decrease esophageal acid exposure. Common therapies include antacids, histamine$_2$-receptor antagonists (H_2RAs), and PPIs. Oral antacids, such as magnesium hydroxide, aluminum hydroxide, and calcium carbonate, will serve to buffer gastric acid and have been shown to be as effective H2-antagonist in treating erosive esophagitis. However, the prolonged use of high-dose antacids has been associated with potential side effects of aluminum toxicity, hypophosphatemic metabolic bone disease, and milk-alkali syndrome.[12,14] Histamine$_2$-receptor antagonists (Famotidine, Ranitidine, Nizatidine, and Cimetidine) selectively block the histamine$_2$-receptor on gastric parietal cells, which reduce the production of gastric acid and pepsin.[28] H_2RAs have been shown to be superior to placebo for improving clinical and endoscopic signs of GERD in adults, but randomized clinical trials showing a similar effect are scarce in pediatrics. In addition, Cimetidine has been associated with gynecomastia and Ranitidine was recently withdrawn from the US market due to the concerns of potential contamination with the probable human carcinogen, *N*-nitrosodimethylamine (NDMA).[29] PPIs irreversible inhibit the hydrogen-potassium-stimulated ATPase pump on gastric parietal cells, leading to a sustained reduction in gastric acid production. PPIs, inclusive of Omeprazole, Lansoprazole, Esomeprazole, and Pantoprazole, have been associated with significant reduction of GERD symptoms in children and are considered superior to H_2RAs and antacids.[30] However, the side effects of PPIs include headache, diarrhea, nausea, constipation, gastric polyposis, as well as potential increased risk for *Clostridium difficile* infection, gastroenteritis, and lower respiratory infection.[12] Alginate, sucralfate, and prokinetic agents (i.e., Erythromycin, Metoclopramide, Bethanecol, and Domperidone) are not routinely used to treat pediatric GERD.[12] Antireflux surgeries (i.e., Nissen fundoplication and Thal fundoplication) have been effective to reduce symptoms of GERD in medically refractory children and adults.

Impact of obesity

The detrimental impact of obesity on the rising prevalence of GERD in adults has been well established and there is growing concern of similar negative health implications for obese children.[3] Obesity is the result of an imbalance of excessive caloric intake with insufficient caloric expenditure and is largely driven by abdominal adiposity. In adults, obesity is defined as a body mass index (BMI) greater than 30 kg/m^2.[31] In pediatrics, obesity is determined by a BMI greater than 95% of that expected for the respective age, gender, weight, and height of the child (see Table 2.1).[32,33] According to the National Center for Health Statistics, the US prevalence of adult obesity increased from 30.5% to 42.4% from 1999 through 2018, while that of children increased from 13.9% in 1999 to 19.3% in 2018.[1,32] This rise in prevalence of obesity demonstrates a ~40% increase among adults and children over the past 20 years and has been noted to affect various sectors of the population differently. Since 1999 obesity has similarly risen among children 6–11 years of age and

Table 2.1 Obesity classification.

Category	Children 2–19 years BMI percentile	Adults >20 years BMI classification
Overweight/preobesity	85th–95th	25–29.9
Obese	≥95th	Class I: 30–34.9 Class II: 35–39.9
Severe obesity	≥120th of 95th percentile	Class III: ≥40

BMI, body mass index (kg/m^2).
Adapted from Fryar CD, Carroll MD, Afful J. Prevalence of overweight, obesity, and severe obesity among children and adolescents aged 2–19 years: United States, 1963–1965 through 2017–2018. NCHS Health E-Stats 2020; Freedman DS, Khan LK, Serdula MK, Galuska DA, Dietz WH. Trends and correlates of class 3 obesity in the United States from 1990 through 2000. JAMA. 2002;288 (14):1758-1761. doi: 10.1001/jama.288.14.1758. PMID: 12365960.

12–19 years of age without a notable difference between genders. However, when considering ethnicity over the last decade, there is a significant difference in obesity prevalence among Non-Hispanic Caucasian (16.1%), African-American (24.2%), Asian (8.7%), Hispanic (25.6%), and Mexican American (26.9%) children of all ages (see Table 2.2).[32] Specifically, African-American girls, Hispanic boys, and Mexican American boys are all noted to have obesity prevalence rates above 28%, which has risen nearly 40% since 2014 (see Table 2.3).[32] Risk factors for childhood obesity include behavioral factors, maternal smoking during pregnancy, breastfeeding, television viewing, and physical activity; however, overweight status during infancy and socioeconomic status may play a role in the differing obesity prevalence rates between various ethnic groups.[34].

Adult obesity is a well-known risk factor for type 2 diabetes mellitus, osteoarthritis, hypertension, stroke, myocardia infarction, Alzheimer disease, depression, and various cancers.[35] Interestingly some obesity risk factors are reversible as one study demonstrated the risk of type 2 diabetes, hypertension, dyslipidemia, and carotid-artery atherosclerosis among overweight/obese children who became nonobese by adulthood to be similar to those children who were never obese.[36] Obesity also contributes to a

Table 2.2 United States childhood obesity prevalence, 1999–2018, Gender profile.

Year	Gender	Age (years)	Obesity prevalence (%)
1999–2000	All	6–11	15.1
		12–19	14.8
	Boys	6–11	15.8
		12–19	14.8
	Girls	6–11	14.3
		12–19	14.8
2017–18	All	6–11	20.3
		12–19	21.2
	Boys	6–11	21.3
		12–19	22.5
	Girls	6–11	19.2
		12–19	19.9

Adapted from Fryar C.D., Carroll M.D., Afful J. Prevalence of overweight, obesity, and severe obesity among children and adolescents aged 2–19 years: United States, 1963–1965 through 2017–2018. NCHS Health E-Stats. 2020.

Table 2.3 2018 United States childhood obesity prevalence, ethnicity profile.

Gender	Non-Hispanic Caucasian (%)	Non-Hispanic African-American (%)	Non-Hispanic Asian (%)	Hispanic (%)	Mexican American (%)
All	16.1	24.2	8.7	25.6	26.9
Boys	17.4	19.4	12.4	28.1	29.2
Girls	14.8	29.1	5.1	23.0	24.9

Adapted from Fryar C.D., Carroll M.D., Afful J. Prevalence of overweight, obesity, and severe obesity among children and adolescents aged 2–19 years: United States, 1963–1965 through 2017–2018. NCHS Health E-Stats. 2020.

significant number of gastrointestinal disorders, namely, GERD, constipation, fecal incontinence, gallstones, acute pancreatitis, and nonalcoholic fatty liver disease.

Multiple studies have demonstrated a worldwide increased prevalence of GERD in obese adults with odds ratios ranging between 1.33 (United States) and 3.3 (Norway) depending on the severity of obesity, observing prevalence rates as high as 50% for adults with a BMI $\geq$ 30 kg/m^2.[2,3] Comparable studies evaluating the prevalence of GERD in the obese pediatric population are fewer in number. In 2009 Malaty et al. determined the prevalence of GERD among boys with a BMI > 95th percentile to be 24.7% compared to the 16.5% GERD prevalence in girls with similar BMIs.[23] In addition, Koebnick et al. found that moderately obese children (BMI $\geq$ 95th percentile or $\geq$ 30 kg/m^2) had up to 30% higher odds of GERD symptoms while extremely obese children (BMI $\geq$ 1.2 $\times$ 95th percentile or a BMI $\geq$ 35 kg/m^2) had up to 40% higher odds of GERD than their normal-weight counterparts.[37] While several pediatric studies show a direct correlation between obesity and GERD symptoms, Elitsur el al conducted a retrospective review of pediatric endoscopy charts and found no significant correlation between BMI Z-score and GERD findings.[38]

The impact of obesity on pediatric esophageal physiology is largely derived from knowledge attained from the adult population. Various mechanisms of altered esophageal function that contribute to GERD occurrences in obese adults have been described, which include increased intraabdominal pressure, decreased LES tone, increased TRLES, and dysfunctional esophageal motility.[39] Abdominal obesity has been confirmed to increase intraabdominal/intragastric pressure, although abdominal visceral adiposity (rather than BMI) appears to be more closely associated with reflux esophagitis.[40,41] In addition, the presence of a hiatal hernia in obese individuals is more likely to be associated with esophagitis due to an altered pressure gradient along the esophagogastric junction.[42] Gastric distension (particularly in the fundus) is known to stimulate TRLES, which is a major contributor to GERD symptoms and is more common in patients with obesity.[3] Multiple studies have shown an increased prevalence (~ 25%–54%) of esophageal dysmotility in obese adults with associated manometric findings of esophagogastric junction outflow obstruction, ineffective esophageal motility, distal esophageal spasm, hypercontractile esophagus, and achalasia being most commonly found.[43–45] Adiponectin is a peptide secreted by adipocytes with decreased plasma levels noted in obese patients, where decreased plasma levels have also been associated with the presence of Barrett esophagus.[46] Conversely, adipocytes and gastric chief cells also produce leptin, which can induce cellular proliferation and elevated plasma levels have been noted in adult men with Barrett esophagus.[3] Lastly, increased esophageal epithelial permeability has been associated with obesity (see Fig. 2.2).[47,48]

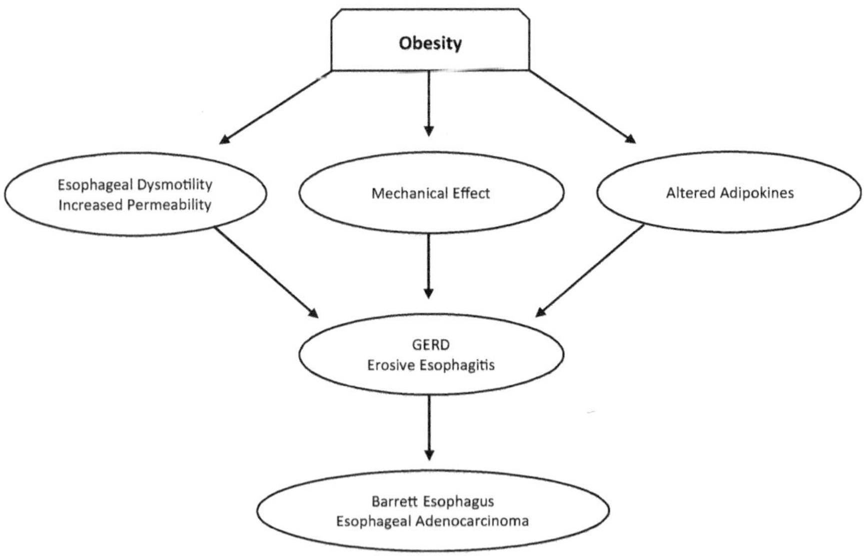

Figure 2.2 Obesity-induced pathophysiology of GERD-related complications. *GERD*, gastroesophageal reflux disease.

Treatment of obesity

Prevention of pediatric obesity is largely dependent on identifying the various risk factors contributing to its onset. Multiple genetic, perinatal, and early-life risk factors have been identified in childhood obesity, including gestational diabetes mellitus, high maternal adiposity, maternal smoking, cesarean delivery, exclusive breastfeeding (protective factor), rapid infantile weight gain, high-caloric food intake during infancy, and early introduction of solid food prior to 6 months of age. Behavioral factors contributing to obesity include increased intake of sweetened beverages and fast food, watching television while eating, fewer family meal times, lower fruit/vegetable intake, and reduced physical activity.[49] Lower socioeconomic status is also associated with increased risk for obesity, likely due to increased food insecurity, unsafe neighborhoods, and greater dependence on unhealthy, calorie-rich foods.[35,50] Obesity prevention strategies for children have targeted the promotion and prescription of healthy dietary and physical activities via elimination of calorie-dense, nutrient-poor foods and practicing 20–60 minutes of vigorous physical activity at least 5 days per week.[51] More recent global prevention strategies are now focusing on community-based/environment-oriented methods that engage not only the individual child, but the family, school, and community.

In 2017 the European Society of Endocrinology and the Pediatric Endocrine Society published new guidelines for the assessment, treatment, and prevention of pediatric obesity.[51] The recommendations for lifestyle intervention focus on reduced consumption of fast foods, sugar-sweetened foods, high-fructose corn syrup, high-fat/ high-sodium/highly processed foods, and saturated fats. They also encourage consumption of whole fruit (rather than fruit juice), increased intake of dietary fiber and vegetables, promotion of regular mealtimes, and avoidance of snacking throughout the day. Physical activities for obesity treatment should include almost daily exercise for 20–60 minutes, as previously mentioned. In addition, these guidelines promote limited nonacademic screen time (<2 hours/day), as well as familial education about healthy food and exercise habits.

Several pharmacotherapeutics to reduce pediatric obesity have recently been approved by the Food and Drug Administration (FDA). Orlistat is a reversible inhibitor of gastric and pancreatic lipases and is FDA approved for use in obese children 12 years and older. Orlistat reduces fat absorption by 30% and can reduce adolescent BMI up to 1.7 kg/m^2 with 6–12 months of use but is also associated with significant gastrointestinal side effects, namely, flatus with discharge, oily stools, frequent defecation, and fecal incontinence.[51,52] Phentermine is FDA approved for adolescents ≥ 16 years old for usage 12 weeks or less and acts to increase catecholamines and serotonin activity in the central nervous system resulting in appetite suppression.[53] Liraglutide, a glucagon-like peptide-1 (GLP-1) agonist, is the latest FDA-approved treatment for obese children 12 years and older: it is mechanism of action is to increase postprandial insulin levels in a glucose-dependent manner, reduce glucagon secretion, delay gastric emptying, and induce weight loss through reductions in appetite and energy intake.[54] While Metformin is not FDA approved for adolescent obesity treatment, it is known to reduce hepatic glucose production, increase peripheral insulin sensitivity, reduce appetite, and is often used to improve glycemic control in teenagers with diabetes mellitus type 2.[51]

Bariatric surgical procedures are considered highly effective long-term solutions for obese adults who have not achieved significant weight loss in response to lifestyle and medical management. Indeed, the performance of bariatric surgery for severely obese adolescents with or without significant comorbidities has doubled within the last two decades.[55] Bariatric surgeries can be classified as restrictive, malabsorptive, or a combination of the two. The Roux-en-Y gastric bypass (RYGB) has long been considered the gold-standard surgery for weight loss in adults and adolescents and is a combination of restrictive and malabsorptive type. During the RYGB, a small gastric pouch is created just distal to the GEJ and is connected directly to the small intestine, bypassing the stomach, duodenum, and the proximal part of the jejunum.[56] The vertical sleeve gastrectomy (VSG) is an increasingly popular restrictive bariatric surgical option where ~85% of the stomach is removed, including the fundus and greater curvature, leaving a narrow gastric lumen.[51] Laparoscopic adjustable gastric banding (LAGB) is a restrictive procedure where a silicone band is placed around the proximal stomach to reduce gastric volume. Unfortunately, the LAGB has been associated with higher complications and is less often performed. In addition to decreasing gastric volume and interfering with intestinal absorption, the RYGB and VSG can decrease the orexigenic hormone ghrelin, while increasing the anorexigenic incretins GLP and peptide YY, thus decreasing appetite and improving insulin resistance.[51] The 3-year efficacy of bariatric surgery in adolescents mirrors that of adults with a notable ~28% weight reduction following the RYGB and ~26% weight reduction after VGB.[55] Secondary comorbidities of adolescent obesity, such as type 2 diabetes, hypertension, and dyslipidemia have also improved following bariatric surgery.[55,57]

Potential complications of bariatric surgery should be considered and include gastrointestinal leaks; incisional hernias; cholelithiasis; small bowel obstruction; malnutrition; vitamin and mineral deficiencies (i.e., iron, vitamin B12, vitamin D, vitamin A, copper, zinc, and selenium); as well as increased suicidal ideation/attempts and worsening mental health status.[51,55] Increased prevalence of GERD, nausea, vomiting, bloating, and diarrhea have also been described following bariatric surgery.[58,59] In a recent policy statement published by the American Academy of Pediatrics pertaining to adolescent bariatric surgery, providers are urged to refer severely obese adolescents to high-quality multidisciplinary bariatric centers with pediatric experts on obesity, adolescent medicine, mental health, nutrition, exercise science, and experienced bariatric surgeons.[57] Contraindications to adolescent bariatric surgery include having a medically correctable cause of obesity, ongoing substance abuse problem, current or planned pregnancy within 12–18 months of surgery, or having a medical, psychiatric, psychosocial, or cognitive condition that prevents postoperative adherence to specified diet and medication regimens.[57]

Complications of obesity-induced gastroesophageal reflux disease

Approximately, 15%–65% of overweight/obese adults seeking weight loss therapy report symptoms of GERD and there is sufficient evidence linking obesity to complications related to long-standing reflux such as erosive esophagitis, Barrett esophagus, and esophageal adenocarcinoma.[3,18,60] Erosive esophagitis is the most common consequence of GERD, occurring in ~25% of adults; and obese adults are 3.3 × as likely to develop erosive esophagitis.[61,62] In children, there is a 12% prevalence of erosive esophagitis irrespective of BMI or weight.[63] Long-term PPI therapy is the first-line treatment for erosive esophagitis in adults and pediatrics; however, antireflux surgery can be considered for medically refractory patients. Peptic strictures due to long-standing esophageal inflammation and fibrosis may occur in 7%–23% of adults and 5%–16% of children with untreated erosive esophagitis and is usually amenable to prolonged PPI therapy and esophageal balloon dilatation.[61,64,65] A known complication of GERD is Barrett esophagus, which is a premalignant esophageal lesion characterized by the metaplastic change of native stratified squamous epithelium to intestinal columnar epithelium. It is estimated that 5.6% of adults in the United States have Barrett esophagus.[61] The exact prevalence of pediatric Barrett esophagus is unknown; however, a 2018 systematic review found that GERD was estimated to be the cause of ~62% of pediatric Barrett esophagus occurrences.[66] The treatment for all Barrett esophagus consists of continuous PPI therapy with intermittent surveillance endoscopies and, in the event of high-grade dysplasia, endoscopic mucosal resection of the Barret segment.[61] The absolute risk of developing esophageal adenocarcinoma is low in nondysplastic Barret esophagus, but this risk increases in the presence of dysplasia. The progression of GERD to erosive esophagitis and further to Barrett esophagus can predispose to the development of esophageal adenocarcinoma typically within the seventh decade of life. Globally, esophageal adenocarcinoma is a low-incidence malignancy (~1.1 cases/100,000 person-years for men and 0.3/100,000 person-years for women); however, the mortality rate is staggering with less than 50% of patients surviving beyond 5 years.[61,67]

Obesity in the form of central adiposity is considered a risk factor for the development of Barrett esophagus and esophageal adenocarcinoma.[68] A *meta*-analysis published in 2008 correlated a 5 kg/m^2 increase in BMI to an increased relative risk for esophageal adenocarcinoma of 1.52 in men and 1.51 in women.[69] Indeed, Hoyo et al. reported an increased odds ratio of 2.39 for esophageal adenocarcinoma in adults with a BMI ≥ 30 kg/m^2, which further increased to 4.76 for adults with BMI ≥ 40 kg/m^2.[70] Data implicating childhood obesity as a significant risk factor for GERD-related complications of Barrett esophagus and esophageal adenocarcinoma are mounting. In 2015 Cook et al. found evidence for a linear and positive association between childhood BMI and future risk of esophageal adenocarcinoma.[71] In addition, Petrick et al. published a prospective cohort study that showed overweight men who were persistently overweight between the ages of 7 and 13 had a risk of developing esophageal adenocarcinoma 3× that of men with normal weight at similar time points. In this same study, men who were overweight at 7 years of age whose weight then normalized by adulthood did not have a significantly increased risk of esophageal adenocarcinoma.[72] The pathophysiological effect of childhood adiposity on esophageal adenocarcinoma may include previously mentioned obesity-induced effects on GERD, namely, direct effect of obesity on GERD, the development of metabolic disorders, android fatty deposition, and hormonal factors.[41,70,72]

Summary

The concurrent rising prevalence of childhood obesity and GERD over the past two decades is not coincidental. Numerous studies have associated various physiological factors of obesity with increased GERD prevalence and the occurrence of GERD-related complications. Early and effective treatment of GERD, especially during childhood, should be encouraged to prevent GERD-related complications in adulthood. The effect of pediatric obesity on impaired esophageal physiology and potential development of malignancy may be reversible with appropriate weight loss therapy. Many lifestyle-based therapeutic options for obesity have proven ineffective for long-term weight loss results; hence, bariatric surgeries have become increasingly popular given the potential to maintain long-term weight loss effects. In addition, the global pandemic of obesity has been recognized by multiple international health organizations who are now advocating for a more aggressive community-based approach to weight loss in addition to individual lifestyle and medicinal weight loss interventions.

References

1. Hales CM, Carroll MD, Fryar CD, Ogden CL. Prevalence of obesity and severe obesity among adults: United States, 2017–2018. NCHS Data Brief, no 360. Hyattsville, MD: National Center for Health Statistics. 2020.
2. El-Serag HB, Sweet S, Winchester CC, Dent J. Update on the epidemiology of gastro-oesophageal reflux disease: a systematic review. *Gut.* 2014;63(6):871–880. Available from: https://doi.org/10.1136/gutjnl-2012-304269. Epub 2013 Jul 13. PMID: 23853213; PMCID: PMC4046948.
3. Chang P, Friedenberg F. Obesity and GERD. *Gastroenterol Clin North Am.* 2014;43(1):161–173. Available from: https://doi.org/10.1016/j.gtc.2013.11.009. Epub 2013 Dec 27. PMID: 24503366; PMCID: PMC3920303.
4. Eusebi LH, Ratnakumaran R, Yuan Y, Solaymani-Dodaran M, Bazzoli F, Ford AC. Global prevalence of, and risk factors for, gastro-oesophageal reflux symptoms: a meta-analysis. *Gut.* 2018;67(3):430–440. Available from: https://doi.org/10.1136/gutjnl-2016-313589. Epub 2017 Feb 23. PMID: 28232473.
5. Zhang Y, Bailey D, Yang P, Kim E, Que J. The development and stem cells of the esophagus. *Development.* 2021;148(6):dev193839. Available from: https://doi.org/10.1242/dev.193839. PMID: 33782045; PMCID: PMC8034879.
6. Cucchiara S, Di Nardo G, Oliva S. Normal motility and development of the esophageal neuroenteric system. In: Kleinman RE, ed. *Walker's Pediatric Gastrointestinal Disease.* People's Medical Publishing House – USA; 2018:59–75.
7. Gold B. Other Esophageal Neuromuscular Disorders. In: Kleinman RE, ed. Walker's Pediatric *Gastrointestinal Disease.* People's Medical Publishing House – USA; 2018:99–123.

8. Argov Z, de Visser M. Dysphagia in adult myopathies. *Neuromuscul Disord.* 2021;31 (1):5–20. Available from: https://doi.org/10.1016/j.nmd.2020.11.001. Epub 2020 Nov 13. PMID: 33334661.
9. Akiyama J, Sumida J, Nakagawa K, et al. New developments in esophageal function testing and esophageal manifestations of connective tissue disorders. *Ann N Y Acad Sci.* 2020;1481(1):170–181. Available from: https://doi.org/10.1111/nyas.14424. Epub 2020 Jul 5. PMID: 32627210.
10. Adams CL, Lohan S, Bruce A, Kamalaraj N, Gunaratne S, White R. Cricopharyngeal bar and dermatomyositis: a cause of rapidly progressive dysphagia. *Int J Rheum Dis.* 2021;24(1):125–131. Available from: https://doi.org/10.1111/1756-185X.14006. Epub 2020 Nov 1. PMID: 33135370.
11. Rosen R, Vandenplas Y, Singendonk M, et al. Pediatric Gastroesophageal Reflux Clinical Practice Guidelines: Joint Recommendations of the North American Society for Pediatric Gastroenterology, Hepatology, and Nutrition and the European Society for Pediatric Gastroenterology, Hepatology, and Nutrition. *J Pediatr Gastroenterol Nutr.* 2018;66(3):516–554. Available from: https://doi.org/10.1097/MPG.0000000000001889. PMID: 29470322; PMCID: PMC5958910.
12. Carroll MW, Jacobson K. Gastroesophageal reflux disease in children and adolescents: when and how to treat. *Paediatr Drugs.* 2012;14(2):79–89. Available from: https://doi.org/10.2165/11594360-000000000-00000.
13. Leung AK, Hon KL. Gastroesophageal reflux in children: an updated review. *Drugs Context.* 2019;8:212591. Available from: https://doi.org/10.7573/dic.212591. PMID: 31258618; PMCID: PMC6586172.
14. Rudolph CD, Hassall E. Gastroesophageal reflux. In: Kleinman RE, ed. *Walker's Pediatric Gastrointestinal Disease.* People's Medical Publishing House – USA; 2018:77–97.
15. Konstantinopoulou S, Sideris GA, DelRosso LM. The role of co-morbidities. *Curr Probl Pediatr Adolesc Health Care.* 2016;46(1):7–10. Available from: https://doi.org/10.1016/j.cppeds.2015.10.010. Epub 2015 Dec 2. PMID: 26655046.
16. Singendonk M, Goudswaard E, Langendam M, et al. Prevalence of gastroesophageal reflux disease symptoms in infants and children: a systematic review. *J Pediatr Gastroenterol Nutr.* 2019;68(6):811–817. Available from: https://doi.org/10.1097/MPG.0000000000002280.
17. Van Howe RS, Storms MR. Gastroesophageal reflux symptoms in infants in a rural population: longitudinal data over the first six months. *BMC Pediatr.* 2010;10:7. Available from: https://doi.org/10.1186/1471-2431-10-7. PMID: 20149255; PMCID: PMC2831886.
18. Okimoto E, Ishimura N, Morito Y, et al. Prevalence of gastroesophageal reflux disease in children, adults, and elderly in the same community. *J Gastroenterol Hepatol.* 2015;30(7):1140–1146. Available from: https://doi.org/10.1111/jgh.12899.
19. Gunasekaran TS, Dahlberg M. Prevalence of gastroesophageal reflux symptoms in adolescents: is there a difference in different racial and ethnic groups? *Dis Esophagus.* 2011;24(1):18–24. Available from: https://doi.org/10.1111/j.1442-2050.2010.01089.x.
20. Chen JH, Wang HY, Lin HH, Wang CC, Wang LY. Prevalence and determinants of gastroesophageal reflux symptoms in adolescents. *J Gastroenterol Hepatol.* 2014;29 (2):269–275. Available from: https://doi.org/10.1111/jgh.12330.
21. Reshetnikov OV, Kurilovich SA, Denisova DV. [Symptoms of gastroesophageal reflux disease and associated factors in adolescents: population survey]. Eksp Klin Gastroenterol. 2013;(12):8-14. Russian. PMID: 24933983.
22. Landau DA, Goldberg A, Levi Z, Levy Y, Niv Y, Bar-Dayan Y. The prevalence of gastrointestinal diseases in Israeli adolescents and its association with body mass index, gender, and Jewish ethnicity. *J Clin Gastroenterol.* 2008;42(8):903–909. Available from: https://doi.org/10.1097/MCG.0b013e31814685f9.

23. Malaty HM, Fraley JK, Abudayyeh S, et al. Obesity and gastroesophageal reflux disease and gastroesophageal reflux symptoms in children. *Clin Exp Gastroenterol.* 2009;2:31–36. Available from: https://doi.org/10.2147/ceg.s4715. Epub 2009 Mar 31. PMID: 21694824; PMCID: PMC3108645.
24. Pashankar DS, Corbin Z, Shah SK, Caprio S. Increased prevalence of gastroesophageal reflux symptoms in obese children evaluated in an academic medical center. *J Clin Gastroenterol.* 2009;43(5):410–413. Available from: https://doi.org/10.1097/MCG.0b013e3181705ce9.
25. Mallah N, Turner JM, González-Barcala FJ, Takkouche B. Gastroesophageal reflux disease and asthma exacerbation: a systematic review and *meta*-analysis. *Pediatr Allergy Immunol.* 2022;33(1):e13655. Available from: https://doi.org/10.1111/pai.13655. Epub 2021 Sep 7. PMID: 34448255.
26. Lightdale JR, Gremse DA. Section on gastroenterology, hepatology, and nutrition. gastroesophageal reflux: management guidance for the pediatrician. *Pediatrics.* 2013;131(5):e1684–e1695. Available from: https://doi.org/10.1542/peds.2013-0421. Epub 2013 Apr 29. PMID: 23629618.
27. Patel NR, Ward MJ, Beneck D, Cunningham-Rundles S, Moon A. The association between childhood overweight and reflux esophagitis. *J Obes.* 2010;2010:136909. Available from: https://doi.org/10.1155/2010/136909. Epub 2010 May 5. PMID: 20700412; PMCID: PMC2911620.
28. van der Pol R, Langendam M, Benninga M, van Wijk M, Tabbers M. Efficacy and safety of histamine-2 receptor antagonists. *JAMA Pediatr.* 2014;168(10):947–954. Available from: https://doi.org/10.1001/jamapediatrics.2014.1273.
29. FDA NEWS RELEASE: FDA Requests Removal of All Ranitidine Products (Zantac) from the Market." http://www.fda.gov, April 1, 2020. http://www.fda.gov/news-events/press-announcements/fda-requests-removal-all-ranitidine-products-zantac-market
30. van der Pol RJ, Smits MJ, van Wijk MP, Omari TI, Tabbers MM, Benninga MA. Efficacy of proton-pump inhibitors in children with gastroesophageal reflux disease: a systematic review. *Pediatrics.* 2011;127(5):925–935. Available from: https://doi.org/10.1542/peds.2010-2719. Epub 2011 Apr 4. PMID: 21464183.
31. Engin A. The definition and prevalence of obesity and metabolic syndrome. *Adv Exp Med Biol.* 2017;960:1–17. Available from: https://doi.org/10.1007/978-3-319-48382-5_1.
32. Fryar CD, Carroll MD, Afful J. Prevalence of overweight, obesity, and severe obesity among children and adolescents aged 2–19 years: United States, 1963–1965 through 2017–2018. *NCHS Health E-Stats.* 2020;.
33. Freedman DS, Khan LK, Serdula MK, Galuska DA, Dietz WH. Trends and correlates of class 3 obesity in the United States from 1990 through 2000. *JAMA.* 2002;288(14):1758–1761. Available from: https://doi.org/10.1001/jama.288.14.1758.
34. Isong IA, Rao SR, Bind M-A, et al. Racial and ethnic disparities in early childhood obesity. *Pediatrics.* 2018;141(1):e20170865.
35. Blüher M. Obesity: global epidemiology and pathogenesis. *Nat Rev Endocrinol.* 2019;15(5):288–298. Available from: https://doi.org/10.1038/s41574-019-0176-8.
36. Juonala M, Magnussen CG, Berenson GS, et al. Childhood adiposity, adult adiposity, and cardiovascular risk factors. *N Engl J Med.* 2011;365(20):1876–1885. Available from: https://doi.org/10.1056/NEJMoa1010112.
37. Koebnick C, Getahun D, Smith N, Porter AH, Der-Sarkissian JK, Jacobsen SJ. Extreme childhood obesity is associated with increased risk for gastroesophageal reflux disease in a large population-based study. *Int J Pediatr Obes.* 2011;6(2-2):e257–e263. Available from: https://doi.org/10.3109/17477166.2010.491118. Epub 2010 Jul 9. PMID: 20615162.

38. Elitsur Y, Dementieva Y, Elitsur R, Rewalt M. Obesity is not a risk factor in children with reflux esophagitis: a retrospective analysis of 738 children. *Metab Syndr Relat Disord.* 2009;7(3):211–214. Available from: https://doi.org/10.1089/met.2008.0069.
39. Taylor SA, Himes R, Hastings E, Garland B. Gastrointestinal conditions in the obese patient. *Adolesc Med State Art Rev.* 2016;27(1):93–108. Spring.
40. Varela JE, Hinojosa M, Nguyen N. Correlations between intra-abdominal pressure and obesity-related co-morbidities. *Surg Obes Relat Dis.* 2009;5(5):524–528. Available from: https://doi.org/10.1016/j.soard.2009.04.003. Epub 2009 Apr 23.
41. Nam SY. Obesity-related digestive diseases and their pathophysiology. *Gut Liver.* 2017;11(3):323–334. Available from: https://doi.org/10.5009/gnl15557. PMID: 27890867; PMCID: PMC5417774.
42. Pandolfino JE, El-Serag HB, Zhang Q, Shah N, Ghosh SK, Kahrilas PJ. Obesity: a challenge to esophagogastric junction integrity. *Gastroenterology.* 2006;130 (3):639–649. Available from: https://doi.org/10.1053/j.gastro.2005.12.016.
43. Suter M, Dorta G, Giusti V, Calmes JM. Gastro-esophageal reflux and esophageal motility disorders in morbidly obese patients. *Obes Surg.* 2004;14(7):959–966. Available from: https://doi.org/10.1381/0960892041719581.
44. Hong D, Khajanchee YS, Pereira N, Lockhart B, Patterson EJ, Swanstrom LL. Manometric abnormalities and gastroesophageal reflux disease in the morbidly obese. *Obes Surg.* 2004;14(6):744–749. Available from: https://doi.org/10.1381/0960892041590854.
45. Popescu AL, Costache RS, Costache DO, et al. Manometric changes of the esophagus in morbidly obese patients. *Exp Ther Med.* 2021;21(6):604. Available from: https://doi.org/10.3892/etm.2021.10036. Epub 2021 Apr 14. PMID: 33936261; PMCID: PMC8082643.
46. Rubenstein JH, Dahlkemper A, Kao JY, et al. A pilot study of the association of low plasma adiponectin and Barrett's esophagus. *Am J Gastroenterol.* 2008;103 (6):1358–1364. Available from: https://doi.org/10.1111/j.1572-0241.2008.01823.x. Epub 2008 May 28. PMID: 18510610.
47. Katzka DA, Kahrilas PJ. Advances in the diagnosis and management of gastroesophageal reflux disease. *BMJ.* 2020;371:m3786. Available from: https://doi.org/10.1136/bmj.m3786.
48. Blevins CH, Dierkhising RA, Geno DM, et al. Obesity and GERD impair esophageal epithelial permeability through 2 distinct mechanisms. *Neurogastroenterol Motil.* 2018;30(10):e13403. Available from: https://doi.org/10.1111/nmo.13403. Epub 2018 Jul 30. PMID: 30062771.
49. Gurnani M, Birken C, Hamilton J. Childhood obesity: causes, consequences, and management. *Pediatr Clin North Am.* 2015;62(4):821–840. Available from: https://doi.org/10.1016/j.pcl.2015.04.001. Epub 2015 May 23. PMID: 26210619.
50. Hemmingsson E. Early childhood obesity risk factors: socioeconomic adversity, family dysfunction, offspring Distress, and junk food self-medication. *Curr Obes Rep.* 2018;7 (2):204–209. Available from: https://doi.org/10.1007/s13679-018-0310-2. PMID: 29704182; PMCID: PMC5958160.
51. Styne DM, Arslanian SA, Connor EL, et al. Pediatric obesity-assessment, treatment, and prevention: an endocrine society clinical practice guideline. *J Clin Endocrinol Metab.* 2017;102(3):709–757. Available from: https://doi.org/10.1210/jc.2016-2573. PMID: 28359099; PMCID: PMC6283429.
52. Czernichow S, Lee CM, Barzi F, et al. Efficacy of weight loss drugs on obesity and cardiovascular risk factors in obese adolescents: a *meta*-analysis of randomized controlled trials. *Obes Rev.* 2010;11(2):150–158. Available from: https://doi.org/10.1111/j.1467-789X.2009.00620.x. Epub 2009 Jul 1. PMID: 19573052.

53. Woodard K, Louque L, Hsia DS. Medications for the treatment of obesity in adolescents. *Ther Adv Endocrinol Metab*. 2020;11. Available from: https://doi.org/10.1177/2042018820918789. PMID: 32523671; PMCID: PMC7257846.
54. Kelly AS, Auerbach P, Barrientos-Perez M, et al. NN8022-4180 Trial Investigators. A randomized, controlled trial of liraglutide for adolescents with obesity. *N Engl J Med*. 2020;382(22):2117–2128. Available from: https://doi.org/10.1056/NEJMoa1916038. Epub 2020 Mar 31. PMID: 32233338.
55. Inge TH, Courcoulas AP, Jenkins TM, et al. Teen-LABS Consortium. Weight loss and health status 3 years after bariatric surgery in adolescents. *N Engl J Med*. 2016;374(2):113–123. Available from: https://doi.org/10.1056/NEJMoa1506699. Epub 2015 Nov 6. PMID: 26544725; PMCID: PMC4810437.
56. Castellana M, Procino F, Biacchi E, et al. Roux-en-Y gastric bypass vs sleeve gastrectomy for remission of type 2 diabetes. *J Clin Endocrinol Metab*. 2021;106(3):922–933. Available from: https://doi.org/10.1210/clinem/dgaa737.
57. Armstrong SC, Bolling CF, Michalsky MP, Reichard KW. SECTION ON OBESITY, SECTION ON SURGERY. Pediatric metabolic and bariatric surgery: evidence, barriers, and best practices. *Pediatrics*. 2019;144(6):e20193223. Available from: https://doi.org/10.1542/peds.2019-3223. Epub 2019 Oct 27. PMID: 31656225.
58. Castilho AVSS, Foratori-Junior GA, Sales-Peres SHC. Bariatric surgery impact on gastroesophageal reflux and dental wear: a systematic review. *Arq Bras Cir Dig*. 2019;32(4):e1466. Available from: https://doi.org/10.1590/0102-672020190001e1466. PMID: 31859919; PMCID: PMC6918764.
59. Dewberry LC, Khoury JC, Ehrlich S, et al. Change in gastrointestinal symptoms over the first 5 years after bariatric surgery in a multicenter cohort of adolescents. *J Pediatr Surg*. 2019;54(6):1220–1225. Available from: https://doi.org/10.1016/j.jpedsurg.2019.02.032. Epub 2019 Feb 28. PMID: 30879757; PMCID: PMC6545240.
60. Garvey WT, Mechanick JI, Brett EM, et al. Reviewers of the AACE/ACE Obesity Clinical Practice Guidelines. American Association of Clinical Endocrinologists and American College of Endocrinology Comprehensive Clinical Practice Guidelines for Medical Care of Patients With Obesity. *Endocr Pract*. 2016;22(Suppl 3):1–203. Available from: https://doi.org/10.4158/EP161365.GL. Epub 2016 May 24. PMID: 27219496.
61. Maret-Ouda J, Markar SR, Lagergren J. Gastroesophageal reflux disease: a review. *JAMA*. 2020;324(24):2536–2547. Available from: https://doi.org/10.1001/jama.2020.21360.
62. Lee HL, Eun CS, Lee OY, et al. Association between GERD-related erosive esophagitis and obesity. *J Clin Gastroenterol*. 2008;42(6):672–675. Available from: https://doi.org/10.1097/MCG.0b013e31806daf64.
63. Gilger MA, El-Serag HB, Gold BD, et al. Prevalence of endoscopic findings of erosive esophagitis in children: a population-based study. *J Pediatr Gastroenterol Nutr*. 2008;47(2):141–146. Available from: https://doi.org/10.1097/MPG.0b013e31815eeabe.
64. Bosheva M, Chatalbashev N, Pechilkova M, Stanev V, Stefanova P. Peptic esophageal stricture in children. *Folia Med (Plovdiv)*. 1998;40(4):24–28.
65. Sag E, Bahadir A, Imamoglu M, et al. Acquired noncaustic esophageal strictures in children. *Clin Exp Pediatr*. 2020;63(11):447–450. Available from: https://doi.org/10.3345/cep.2020.00199. Epub 2020 Oct 15. PMID: 33137248; PMCID: PMC7642133.
66. Raicevic M, Saxena AK. Barrett's esophagus in children: what is the evidence? *World J Pediatr*. 2018;14(4):330–334. Available from: https://doi.org/10.1007/s12519-018-0170-6. Epub 2018 Jul 10. PMID: 29992379.

67. Uhlenhopp DJ, Then EO, Sunkara T, Gaduputi V. Epidemiology of esophageal cancer: update in global trends, etiology and risk factors. *Clin J Gastroenterol.* 2020;13 (6):1010–1021. Available from: https://doi.org/10.1007/s12328-020-01237-x. Epub 2020 Sep 23. PMID: 32965635.
68. Nimptsch K, Steffen A, Pischon T. Obesity and oesophageal cancer. *Recent Results Cancer Res.* 2016;208:67–80. Available from: https://doi.org/10.1007/978-3-319-42542-9_4.
69. Renehan AG, Tyson M, Egger M, Heller RF, Zwahlen M. Body-mass index and incidence of cancer: a systematic review and *meta*-analysis of prospective observational studies. *Lancet.* 2008;371(9612):569–578. Available from: https://doi.org/10.1016/S0140-6736(08)60269-X.
70 Hoyo C, Cook MB, Kamangar F, et al. Body mass index in relation to oesophageal and oesophagogastric junction adenocarcinomas: a pooled analysis from the International BEACON Consortium. *Int J Epidemiol.* 2012;41(6):1706–1718. Available from: https://doi.org/10.1093/ije/dys176. Epub 2012 Nov 12. PMID: 23148106; PMCID: PMC3535758.
71. Cook MB, Freedman ND, Gamborg M, Sørensen TI, Baker JL. Childhood body mass index in relation to future risk of oesophageal adenocarcinoma. *Br J Cancer.* 2015;112(3):601–607. Available from: https://doi.org/10.1038/bjc.2014.646. Epub 2015 Jan 6. PMID: 25562436; PMCID: PMC4453659.
72. Petrick JL, Jensen BW, Sørensen TIA, Cook MB, Baker JL. Overweight patterns between childhood and early adulthood and esophageal and gastric cardia adenocarcinoma risk. *Obes (Silver Spring).* 2019;27(9):1520–1526. Available from: https://doi.org/10.1002/oby.22570. Epub 2019 Aug 5. PMID: 31380608; PMCID: PMC6707875.

CHAPTER 3

Obesity and impact on gastroesophageal reflux disease

Akinari Sawada, Ilia Sergeev and Daniel Sifrim
Wingate Institute of Neurogastroenterology, Blizard Institute, Barts and The London School of Medicine and Dentistry, Queen Mary University of London, London, United Kingdom

Structural and physiological changes associated with obesity that contributes to gastroesophageal reflux disease

Gastroesophageal reflux disease (GERD) is often seen in obese patients. Many studies reported a positive association between GERD and obesity.[1–11] A recent *meta*-analysis showed that obesity (BMI $\geq$ 30 kg/m^2) is one of the risk factors for GERD together with age, smoking, and nonsteroidal antiinflammation drug/aspirin use. Its odds ratio was 1.73 (95% CI: 1.46–2.06) compared to nonobese individuals. A dose-dependent association was shown between reflux symptoms and BMI (i.e., larger BMI more frequent or severe reflux symptoms),[1–4,10] and two longitudinal studies found that weight gain during approximately 10 years increased the risk of developing reflux symptoms.[3,8] Complications of GERD such as reflux esophagitis, Barrett's esophagus (BE), and esophageal adenocarcinoma are also related to obesity.[7,10,12] The mechanisms by which obesity causes GERD can be explained by its effects on structural and physiological functions of the gastroesophageal antireflux barrier. Fig. 3.1 shows the obesity-related changes contributing to GERD, which will be discussed next.

Structural changes associated with gastroesophageal reflux disease in obesity

Hiatal hernia (HH) is a condition in which the contents of the abdominal cavity herniated into the mediastinum through the diaphragmatic hiatus. HH is one of the most important structural changes contributing to acid reflux. Sliding hernia is the predominant type in obesity while paraesophageal hernia accounts for a lower proportion of cases (<10%).[13,14]

Obesity and Esophageal Disorders
DOI: https://doi.org/10.1016/B978-0-323-98365-5.00005-X

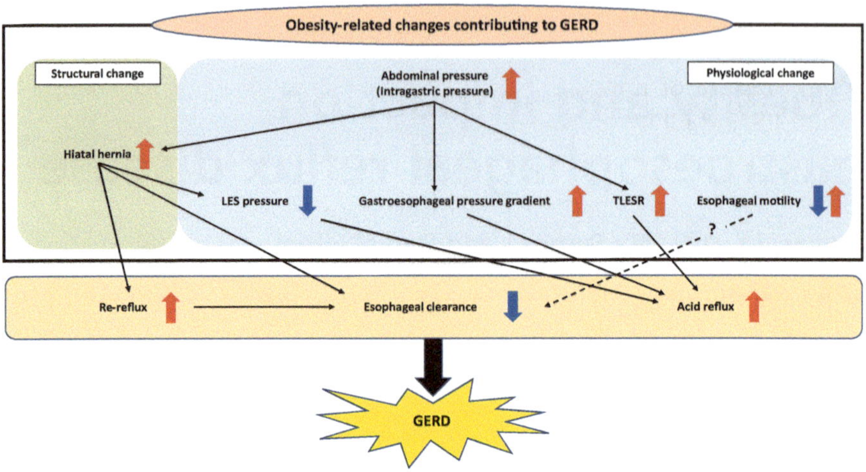

Figure 3.1 Structural and physiological changes in obesity leading to changes in esophageal clearance and contributing to GERD. *GERD*, Gastroesophageal reflux disease.

Several studies showed an increased prevalence of HH (up to 52.6%) in obese patients.[15–27] In a *meta*-analysis, preoperative esophagogastroduodenoscopy (EGD) found a prevalence of 19.7% of HH in 4511 obese patients with a mean BMI of 47 kg/m^2.[28] Hiatal hernias larger than 2 cm are found in approximately 10% of obese HH patients.[18,24] Several risk factors of HH have been identified such as older age,[25] Black (compared to Caucasians and Hispanics),[25] reflux symptoms,[19,26] and female.[26]

Pandolfino et al.[29] showed that the spatial separation between the lower esophageal sphincter (LES) and the crural diaphragm (CD) increases in overweight and obese patients compared to subjects with normal BMI. Moreover, the LES-CD separation significantly correlated to BMI and waist circumstance. Lee et al. demonstrated that abdominal compression with a waist belt displaces the squamocolumnar junction proximally as seen in subjects with increased waist circumference.[30] These studies implicate a causal relationship between obesity and HH.

Physiological changes associated with gastroesophageal reflux disease in obesity

Lower esophageal sphincter pressure

Obese patients often show lower LES pressure than healthy subjects; however, the LES pressure does not decrease inversely proportional to BMI.[31–34] It is probably due to concomitant HH rather than the effect of sheer obesity on the LES.[35,36] The reduction of LES pressure in HH is attributable to the LES-CD separation because esophagogastric junction (EGJ) pressure of normal subjects does not differ from the sum of separated LES and CD pressures in HH.[37] Interestingly, increased intraabdominal pressure with waist belt increases the LES pressure in both healthy volunteers and GERD patients.[30,38–40] Mittal et al. concludes that tonic contraction of the crural diaphragm contributes to this LES response.[39]

Transient lower esophageal sphincter relaxation

Obesity impacts the major mechanism of gastroesophageal reflux, that is, transient lower esophageal sphincter relaxations (TLESR).[35,41] Wu et al. evaluated the effect of obesity on TLESR.[35] Obese patients showed more postprandial TLESRs than normals after ingestion of standardized test meal. More importantly, TLESRs in obese patients were more often accompanied by acid refluxes. The total number of TLESRs significantly correlated to BMI as well as waist circumstance.

Intragastric pressure and gastroesophageal pressure gradient

Intragastric pressure has a positive correlation with BMI and waist circumference. Therefore higher intragastric pressure in obesity results in higher gastroesophageal pressure gradient compared to normal-weight subjects.[29,30,40] The abdominal fat seems to create high intraabdominal pressure because abdominal compression using a waist belt can mimic the pressure change happening in patients with large waist circumference (i.e., central obesity).[30] The experimental abdominal compression in reflux patients increased esophageal acid exposure postprandially via impaired acid clearance and increased TLESRs with acid reflux.[38] Re-reflux events underlie poor esophageal clearance.

Esophageal motility

Both hyper- and hypomotility disorders can be seen in up to 54% of obese patients.[34,42,43] Obesity seems to have little effect on the contraction amplitude of the esophageal body.[32,33,36,44] Interestingly, majority of obese patients with esophageal motility disorders are asymptomatic.[34,42]

The impact of obesity on swallow-induced LES relaxation is controversial. Rogers et al.[36] demonstrated a higher 4-second integrated relaxation pressure and a higher proportion of EGJ outflow obstruction in nonobese GERD patients compared to obese, whereas Yen et al. found the opposite.[45]

Does obesity phenotype matter to gastroesophageal reflux disease/dyspepsia and role of genomic testing?

Impact of body fat distribution in gastroesophageal reflux disease

Obesity is generally categorized on the basis of BMI but BMI reflects body fat content moderately, and cannot provide the detail of either body composition or fat distribution. A different fat distribution is likely to matter to the pathophysiology of GERD given the male-predominant nature of esophageal adenocarcinoma.[46] Several studies found that fat distribution plays a more important role in GERD than BMI.[47–50] Body fat distribution can be measured anthropometrically such as waist circumference and waist–hip ratio. Besides, computed tomography can visualize and calculate the area of abdominal visceral adipose tissue. A *meta*-analysis by Singh et al. showed that central obesity increased the risk of erosive esophagitis (EE) (aOR, 1.93; 95% CI: 1.38–2.71), BE (aOR, 1.88; 95% CI: 1.20–2.95) even after the adjustment for BMI.[12] Corley et al. reported the positive association between abdominal diameter and reflux symptom independent of BMI. However, its effect was disproportional depending on race and gender.[50] This is consistent with a Japanese study showing the negative association between central adiposity and EE.[11]

There are some plausible mechanisms underlying the association between central adiposity and GERD. A physiology study revealed that waist circumference has a more direct effect on intragastric pressure and gastroesophageal pressure gradient than BMI.[29] Furthermore, the abdominal fat visceral adipose tissue creates insulin resistance by releasing free fatty acids, tumor necrosis factor α (TNF-α), and resistin. A subsequent increase in insulin and insulin growth factor 1 might lead to carcinogenesis.[51] Additionally, proinflammatory cytokines such as TNF-α, interleukin-6, adiponectin, and leptin could promote esophageal inflammation and metaplasia even though there is still controversy about such association.[52–56]

Genes associated with obesity and gastroesophageal reflux disease

One twin study revealed that monozygotic twins develop GERD more likely than dizygotic twins with an estimated heritability of 43%.[57] Another study found that genetic factors override shared environmental effects in terms of GERD.[58] These studies indicate that genetic factor matters to GERD.

A recent genome-wide association study (GWAS) using a large number of cases identified single nucleotide polymorphisms (SNPs) (rs491603 at chromosome 1p34.3), which interact with the association between obesity and BE. Obese individuals with rs491603-AA genotype showed a 3.3-fold higher risk of BE compared to nonobese, whereas the risk of BE was elevated by just 1.52-fold in obesity with rs491603-GG genotype than nonobese.[59] It should be noted that the interaction did not reach genome-wide significance (i.e., $P < 5 \times 10^{-8}$) and other previous GWAS with relatively small sample size found inconsistent SNPs associated with the interaction between BMI and GERD.[60,61] If we can identify the SNPs related to the higher risk of EE, BE, or EAC (esophageal adenocarcinoma), it would make the surveillance program much more efficient.

Diagnostic modalities, pathophysiology, and management options

The diagnosis of GERD is based on combination of clinical, endoscopic, and physiologic parameters, most recently discussed by the 2018 Lyon Consensus.[62] Neither clinical symptoms alone nor an empiric proton pump inhibitor (PPI) trial is sensitive or specific enough for GERD diagnosis.[63,64]

Endoscopic techniques

Conventional endoscopy alone lacks the sensitivity and specificity for the diagnosis of GERD. Definite endoscopic diagnosis of GERD can be made in the presence of high-grade esophagitis (LA grade C&D) or the presence of long-segment Barrett's mucosa. However, the presence of low-grade (LA A&B) esophagitis is a nonspecific finding, present in up to 7.5% of asymptomatic patients.[65] Moreover, up to 70% of patients with GERD symptoms have normal-appearing mucosa on endoscopy,[66] and additional diagnostic modalities are needed to differentiate between GERD, NERD (non erosive reflux disease), and functional dyspepsia.

The use of image-enhanced endoscopy for the diagnosis of GERD was previously studied. The presence of an increased number of intrapapillary capillary loops, microerosions, and a nonround pit pattern below the squamocolumnar junction on narrow-band imaging (NBI) was all associated with the diagnosis of GERD.[67–69] Likewise, dilatation of intercellular space and increased number of intrapapillary capillary loops as seen using confocal laser endomicroscopy (CLE) were associated with

GERD diagnosis.[70] However, currently the previously mentioned techniques are either not widely available (CLE) or not routinely used (NBI) for diagnosis of reflux outside the clinical research studies.

Mucosal impedance testing (MIT) is a technique that allows real-time mucosal impedance evaluation using a through-the-scope impedance probe. The mucosal impedance values correlate with histological findings of epithelial barrier dysfunction, potentially allowing the diagnosis of GERD during the endoscopic procedure.[71] Additionally, previously published studies demonstrated the ability of MIT to discriminate between GERD and non-GERD conditions.[72–74]

Endoscopic pressure study integrated system (EPSIS) is a technique that evaluates the lower esophageal sphincter (LES) function. During the EGD, the stomach is insufflated with gas, and gastric pressure is continuously monitored. A correlation between intragastric pressure measured using EPSIS and the diagnosis of GERD on 24-hour pH monitoring was demonstrated in several studies.[75,76]

Reflux monitoring

Multichannel intraluminal impedance (MII) monitoring, coupled with pH-metry, is considered a "gold standard" for GERD diagnosis. During the examination, a trans-nasal catheter with pH and impedance sensors is positioned in the patient's esophagus, and a recording is being carried out for approximately 24 hours, while the patient continues with his normal daily routine. Fig. 3.2 shows MII in an obese patient, specifically a patient with a hiatal hernia, to illustrate the positioning of the device. Documentation of Acid Exposure Time (AET) $>6\%$ is diagnostic of GERD, and AET $<4\%$ and less than 40 reflux episodes rules out the GERD diagnosis.[62]

Ambulatory catheter-based pH-metry provides 24-hour recording of lower esophageal acid exposure, allowing measurement of AET. However, the lack of impedance monitoring prevents nonacid reflux detection and hampers a correct analysis of reflux and symptom association.

To avoid prolonged nasal discomfort and day-to-day reflux variability, prolonged esophageal pH-metry can be monitored using a wireless pH capsule. The capsule is usually positioned in the lower esophagus during the EGD and allows ambulatory pH measurement and recording for a

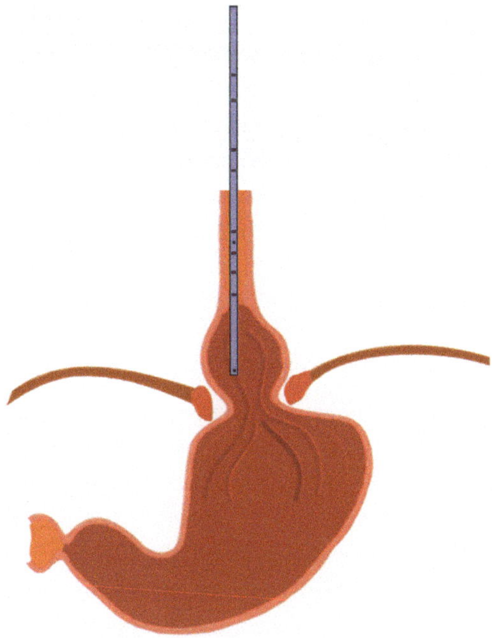

Figure 3.2 Impedance pH-metry in an obese patient with hiatal hernia.

period of 48–72 hours. It allows to study the acid exposure both without (the first 24 hours) and subsequently with the PPI treatment. Similar to ambulatory catheter-based pH-metry, the main disadvantage of the technique is the inability to detect nonacid and/or gas reflux episodes, making it difficult to assess the correlation between symptoms and these reflux episodes.

Other diagnostic techniques

The use of schintigraphy for GERD diagnosis was abandoned in the past due to insufficient sensitivity and low reproducibility. However, the use of current technologies and updated study protocols in esophageal and extra-esophageal manifestations of GERD[77–79] demonstrated a good correlation to 24-hours pH-metry, making the scintigraphy a potentially noninvasive diagnostic tool in GERD patients.

Measurement of salivary pepsin concentration is being evaluated as a noninvasive test of GERD, which might be of great advantage for both adult and pediatric populations. Matsumura et al. found increased pepsin levels in patients with AET > 6%, as compared to those with AET < 6%,[80] and Li et al. have demonstrated higher levels of pepsin in the saliva of patients with nonerosive esophagitis compared to those with functional dyspepsia.[81] In pediatric patients, Haddad et al. found that pepsin levels were associated with reflux episodes and peak values correlated with acidic reflux.[82] However, it appears that current pepsin tests lack the necessary sensitivity and specificity and are not yet ready for clinical use.[83,84]

Gastroesophageal reflux disease treatment

Treatment of GERD consists of lifestyle modification, pharmacologic, endoscopic, and surgical interventions, or a combination of the above. The common therapeutic goal is to control symptoms and achieve the resolution of significant esophagitis, if present.

Lifestyle modification's role in alleviating GERD symptoms is well studied. High-fat meals, alcohol, and carbonated drinks have all been associated with increased GERD symptoms,[85] and Mediterranean diet was associated with lower risk of GERD symptoms.[86] Likewise, increasing interval between last meal and sleep, avoidance of late snacks, and head of the bed elevation could all decrease the GERD probability.[87] Last but not least, tobacco cessation, in addition to other obvious health benefits, is associated with the improvement of GERD symptoms and should be encouraged in all smokers.

PPI-type drugs constitute the backbone of pharmacological GERD treatment, aiming to decrease gastric acidity. This treatment is effective, well tolerated, and with paucity of serious side effects.[88] As with any medication, minimal effective dosage should be sought, and the treatment should be discontinued if no longer needed.

In patients with GERD symptoms refractory to PPI, an addition of Histamine receptor blockers (H2 blockers) might be to further suppress gastric acid secretion. This combination is especially effective for nighttime symptom suppression.[89] The use of alginates or mucosal protective agents might also be used as an add-on to PPI in severe cases or as a solitary therapy in mild cases.

Potassium-competitive acid blockers are new generation acid suppression drugs that were shown to have a rapid onset of action and to be highly effective in the alleviation of GERD symptoms and treatment of EE.[90]

However, this type of drug is currently only licensed in several countries in the East and is unavailable in the majority of western countries.

Endoscopic and surgical therapies with or without bariatric interventions will be discussed at length elsewhere in this book.

Challenges in the treatment of gastroesophageal reflux disease in obesity

Diaphragmatic hernia is a common finding in obese patients, with prevalence ranging from 18% to almost 40% in different cohorts.[24,25,45,91] The presence of a hiatal hernia disrupts the esophageal antireflux barrier mechanism, and the hernia sack acts as a reservoir that allows acid backflow during swallowing. Furthermore, hiatal hernia is associated with impaired esophageal reflux clearance,[92,93] higher acid exposure,[93,94] increased symptoms of reflux, and higher prevalence and severity of reflux esophagitis.[95,96] Likewise, hiatal hernia is associated with reduced efficacy of PPI.[97] Presence of large hiatal hernia in obese patients, especially if potential candidate to bariatric surgery, should warrant antireflux surgery consideration. In the light of the above, The ICARUS guidelines suggest that patients with GERD symptoms and an endoscopic diagnosis of a hiatal hernia might be good candidates for antireflux surgery.[95]

The question whether obesity by itself is associated with a reduced response to PPI treatment is a matter of controversy. It was previously suggested that due to changes in distribution volume, metabolism, and elimination, obesity might alter the pharmacokinetics of PPI in obese.[98–100] A *meta*-analysis performed by Sharma et al.[101] found that higher BMI was associated with more severe baseline EE, and higher grades of EE were associated with lower probability of EE healing. However, both the EE healing and the GERD symptoms resolution were similar across all BMIs. Therefore it appears that obesity alone, without the presence of additional factors such as hiatal hernia, is not associated with PPI refractoriness.

Past studies demonstrated that significant reflux disease and EE can be present in obese patients without symptoms.[102–104] Moreover, reduced sensitivity of morbidly obese patients to direct acid exposure of esophageal mucosa was demonstrated by Ortiz et al.[105] As, in clinical practice, GERD treatment strategies are usually driven by symptoms, a significant proportion of obese patients will not be diagnosed with GERD and therefore left without treatment. Although screening for esophageal conditioned such as Barret's in asymptomatic obese patients is not routinely recommended at this point,[106] EGD is recommended as a part of an evaluation for patients planned for bariatric surgery.[103,107]

Bariatric surgery can alleviate GERD by several mechanisms. In addition to weight reduction, which by itself reduces the reflux symptoms, anatomic alterations after the surgery allow for reduced reflux. Thus Roux-en-Y gastric bypass excludes the esophagus from both the acidic and bile reflux and is very effective both as bariatric and antireflux procedure.[108,109] Likewise, LES augmentation through fundoplication in addition to bariatric surgery is effective in reflux control.[110]

Several endoscopic GERD interventions are currently in clinical use or under investigation.[111] Probably the most studied is the use of radiofrequency to deliver thermal energy to gastric cardia and lower esophagus (Stretta therapy). A procedure is performed under sedation in outpatient setting and results in subjective and objective improvements.[112,113] Endoscopic fundoplication techniques and devices, such as transoral incisionless fundoplication or medigus ultrasonic surgical endostapler, are in use and will be discussed in detail elsewhere.

References

1. Murray L, Johnston B, Lane A, et al. Relationship between body mass and gastro-oesophageal reflux symptoms: the Bristol Helicobacter Project. *Int J Epidemiol.* 2003; 32(4):645–650.
2. Locke GR, Talley NJ, Fett SL, Zinsmeister AR, Melton LJ. Risk factors associated with symptoms of gastroesophageal reflux. *Am J Med.* 1999;106(6):642–649.
3. Nilsson M, Johnsen R, Ye W, Hveem K, Lagergren J. Obesity and estrogen as risk factors for gastroesophageal reflux symptoms. *JAMA.* 2003;290(1):66–72.
4. Ponce J, Vegazo O, Beltrán B, et al. Prevalence of gastro-oesophageal reflux disease in Spain and associated factors. *Aliment Pharmacol Ther.* 2006;23(1):175–184.
5. Diaz-Rubio M, Moreno-Elola-Olaso C, Rey E, Locke GR, Rodriguez-Artalejo F. Symptoms of gastro-oesophageal reflux: prevalence, severity, duration and associated factors in a Spanish population. *Aliment Pharmacol Ther.* 2004;19(1):95–105.
6. Wu AH, Tseng CC, Bernstein L. Hiatal hernia, reflux symptoms, body size, and risk of esophageal and gastric adenocarcinoma. *Cancer.* 2003;98(5):940–948.
7. El-Serag HB, Graham DY, Satia JA, Rabeneck L. Obesity is an independent risk factor for GERD symptoms and erosive esophagitis. *Am J Gastroenterol.* 2005; 100(6):1243–1250.
8. Jacobson BC, Somers SC, Fuchs CS, Kelly CP, Camargo CA. Body-mass index and symptoms of gastroesophageal reflux in women. *N Engl J Med.* 2006; 354(22):2340–2348.
9. Hampel H, Abraham NS, El-Serag HB. *Meta*-analysis: obesity and the risk for gastroesophageal reflux disease and its complications. *Ann Intern Med.* 2005; 143(3):199–211.

10. Nocon M, Labenz J, Jaspersen D, et al. Association of body mass index with heartburn, regurgitation and esophagitis: results of the progression of gastroesophageal reflux disease study. *J Gastroenterol Hepatol*. 2007;22(11):1728–1731.
11. Gunji T, Sato H, Iijima K, et al. Risk factors for erosive esophagitis: a cross-sectional study of a large number of Japanese males. *J Gastroenterol*. 2011;46(4):448–455.
12. Singh S, Sharma AN, Murad MH, et al. Central adiposity is associated with increased risk of esophageal inflammation, metaplasia, and adenocarcinoma: a systematic review and *meta*-analysis. *Clin Gastroenterol Hepatol*. 2013;11(11):1399–1412. e1397.
13. Shada AL, Stem M, Funk LM, Greenberg JA, Lidor AO. Concurrent bariatric surgery and paraesophageal hernia repair: comparison of sleeve gastrectomy and Roux-en-Y gastric bypass. *Surg Obes Relat Dis*. 2018;14(1):8–13.
14. Boules M, Corcelles R, Guerron AD, et al. The incidence of hiatal hernia and technical feasibility of repair during bariatric surgery. *Surgery*. 2015;158(4):911–916. discussion 916–918.
15. Suter M, Dorta G, Giusti V, Calmes JM. Gastro-esophageal reflux and esophageal motility disorders in morbidly obese patients. *Obes Surg*. 2004;14(7):959–966.
16. Küper MA, Kratt T, Kramer KM, et al. Effort, safety, and findings of routine preoperative endoscopic evaluation of morbidly obese patients undergoing bariatric surgery. *Surg Endosc*. 2010;24(8):1996–2001.
17. Ricci G, Amella C, Forti E, et al. 24-h pH-metry and multichannel intraluminal impedance monitoring in obese patients with and without gastroesophageal reflux disease symptoms. *Obes Surg*. 2011;21(1):48–53.
18. Ozeki KA, Tran SA, Cheung R, Eisenberg D. Preoperative endoscopic findings in veterans undergoing bariatric surgery: prevalence and predictors of Barrett's Esophagus. *Obes Surg*. 2020;30(2):657–663.
19. Ghaderi I, Gondal AB, Samamé J, Serrot F, Galvani CA. Preoperative endoscopic and radiologic evaluation of bariatric patients: what do they add? *J Gastrointest Surg*. 2020;24(4):764–771.
20. Heimgartner B, Herzig M, Borbély Y, Kröll D, Nett P, Tutuian R. Symptoms, endoscopic findings and reflux monitoring results in candidates for bariatric surgery. *Dig Liver Dis*. 2017;49(7):750–756.
21. de Moura Almeida A, Cotrim HP, Santos AS, et al. Preoperative upper gastrointestinal endoscopy in obese patients undergoing bariatric surgery: is it necessary? *Surg Obes Relat Dis*. 2008;4(2):144–149. discussion 150-141.
22. Mohammed R, Fei P, Phu J, Asai M, Antanavicius G. Efficiency of preoperative esophagogastroduodenoscopy in identifying operable hiatal hernia for bariatric surgery patients. *Surg Obes Relat Dis*. 2017;13(2):287–290.
23. Salama A, Saafan T, El Ansari W, Karam M, Bashah M. Is routine preoperative esophagogastroduodenoscopy screening necessary prior to laparoscopic sleeve gastrectomy? Review of 1555 cases and comparison with current literature. *Obes Surg*. 2018;28(1):52–60.
24. Che F, Nguyen B, Cohen A, Nguyen NT. Prevalence of hiatal hernia in the morbidly obese. *Surg Obes Relat Dis*. 2013;9(6):920–924.
25. Mani VR, Kalabin A, Nwakanama C, Suman P, Ahmed L. Preoperative vs intraoperative diagnosis of hiatal hernia in bariatric population. *Surg Obes Relat Dis*. 2019; 15(11):1949–1955.
26. Goitein D, Sakran N, Rayman S, Szold A, Goitein O, Raziel A. Barium swallow for hiatal hernia detection is unnecessary prior to primary sleeve gastrectomy. *Surg Obes Relat Dis*. 2017;13(2):138–142.
27. Mizrahi I, Abubeih A, Rachmuth J, et al. Routine upper gastrointestinal fluoroscopy before laparoscopic sleeve gastrectomy: Is it necessary? *Obes Surg*. 2019; 29(6):1704–1708.

28. Parikh M, Liu J, Vieira D, et al. Preoperative endoscopy prior to bariatric surgery: a systematic review and *meta*-analysis of the literature. *Obes Surg*. 2016; 26(12):2961–2966.
29. Pandolfino JE, El-Serag HB, Zhang Q, Shah N, Ghosh SK, Kahrilas PJ. Obesity: a challenge to esophagogastric junction integrity. *Gastroenterology*. 2006;130(3): 639–649.
30. Lee YY, Wirz AA, Whiting JG, et al. Waist belt and central obesity cause partial hiatus hernia and short-segment acid reflux in asymptomatic volunteers. *Gut*. 2014; 63(7):1053–1060.
31. Schneider JM, Brücher BL, Küper M, Saemann K, Königsrainer A, Schneider JH. Multichannel intraluminal impedance measurement of gastroesophageal reflux in patients with different stages of morbid obesity. *Obes Surg*. 2009;19(11):1522–1529.
32. Tolone S, Limongelli P, del Genio G, et al. Gastroesophageal reflux disease and obesity: do we need to perform reflux testing in all candidates to bariatric surgery? *Int J Surg*. 2014;12(suppl 1):S173–S177.
33. Küper MA, Kramer KM, Kirschniak A, et al. Dysfunction of the lower esophageal sphincter and dysmotility of the tubular esophagus in morbidly obese patients. *Obes Surg*. 2009;19(8):1143–1149.
34. Jaffin BW, Knoepflmacher P, Greenstein R. High prevalence of asymptomatic esophageal motility disorders among morbidly obese patients. *Obes Surg*. 1999;9(4):390–395.
35. Wu JC, Mui LM, Cheung CM, Chan Y, Sung JJ. Obesity is associated with increased transient lower esophageal sphincter relaxation. *Gastroenterology*. 2007; 132(3):883–889.
36. Rogers BD, Patel A, Wang D, Sayuk GS, Gyawali CP. Higher esophageal symptom burden in obese subjects results from increased esophageal acid exposure and not from dysmotility. *Clin Gastroenterol Hepatol*. 2020;18(8):1719–1726.
37. Kahrilas PJ, Lin S, Chen J, Manka M. The effect of hiatus hernia on gastro-oesophageal junction pressure. *Gut*. 1999;44(4):476–482.
38. Mitchell DR, Derakhshan MH, Wirz AA, Ballantyne SA, McColl KEL. Abdominal compression by waist belt aggravates gastroesophageal reflux, primarily by impairing esophageal clearance. *Gastroenterology*. 2017;152(8):1881–1888.
39. Mittal RK, Fisher M, McCallum RW, Rochester DF, Dent J, Sluss J. Human lower esophageal sphincter pressure response to increased intra-abdominal pressure. *Am J Physiol*. 1990;258(4 Pt 1):G624–G630.
40. Derakhshan MH, Robertson EV, Fletcher J, et al. Mechanism of association between BMI and dysfunction of the gastro-oesophageal barrier in patients with normal endoscopy. *Gut*. 2012;61(3):337–343.
41. Schneider JH, Küper M, Königsrainer A, Brücher B. Transient lower esophageal sphincter relaxation in morbid obesity. *Obes Surg*. 2009;19(5):595–600.
42. Koppman JS, Poggi L, Szomstein S, Ukleja A, Botoman A, Rosenthal R. Esophageal motility disorders in the morbidly obese population. *Surg Endosc*. 2007;21(5):761–764.
43. Hong D, Khajanchee YS, Pereira N, Lockhart B, Patterson EJ, Swanstrom LL. Manometric abnormalities and gastroesophageal reflux disease in the morbidly obese. *Obes Surg*. 2004;14(6):744–749.
44. Anggiansah R, Sweis R, Anggiansah A, Wong T, Cooper D, Fox M. The effects of obesity on oesophageal function, acid exposure and the symptoms of gastro-oesophageal reflux disease. *Aliment Pharmacol Ther*. 2013;37(5):555–563.
45. Yen HH, Tseng PH, Shih MC, Yang PJ, Lin MT, Lee PC. Derangement of esophageal anatomy and motility in morbidly obese patients: a prospective study based on high-resolution impedance manometry. *Surg Obes Relat Dis*. 2020;16(12):2006–2015.
46. Coleman HG, Xie SH, Lagergren J. The epidemiology of esophageal adenocarcinoma. *Gastroenterology*. 2018;154(2):390–405.

47. Corley DA, Kubo A, Levin TR, et al. Abdominal obesity and body mass index as risk factors for Barrett's esophagus. *Gastroenterology*. 2007;133(1):34–41.
48. Edelstein ZR, Farrow DC, Bronner MP, Rosen SN, Vaughan TL. Central adiposity and risk of Barrett's esophagus. *Gastroenterology*. 2007;133(2):403–411.
49. Nam SY, Choi IJ, Ryu KH, Park BJ, Kim HB, Nam BH. Abdominal visceral adipose tissue volume is associated with increased risk of erosive esophagitis in men and women. *Gastroenterology*. 2010;139(6):1902–1911. e1902.
50. Corley DA, Kubo A, Zhao W. Abdominal obesity, ethnicity and gastro-oesophageal reflux symptoms. *Gut*. 2007;56(6):756–762.
51. Greer KB, Thompson CL, Brenner L, et al. Association of insulin and insulin-like growth factors with Barrett's oesophagus. *Gut*. 2012;61(5):665–672.
52. Rubenstein JH, Morgenstern H, McConell D, et al. Associations of diabetes mellitus, insulin, leptin, and ghrelin with gastroesophageal reflux and Barrett's esophagus. *Gastroenterology*. 2013;145(6):1237–1244. e1231–1235.
53. Rubenstein JH, Kao JY, Madanick RD, et al. Association of adiponectin multimers with Barrett's oesophagus. *Gut*. 2009;58(12):1583–1589.
54. Chandar AK, Devanna S, Lu C, et al. Association of serum levels of adipokines and insulin with risk of Barrett's esophagus: a systematic review and *meta*-analysis. *Clin Gastroenterol Hepatol*. 2015;13(13):2241–2255. e2241–2244; quiz e2179.
55. Francois F, Roper J, Goodman AJ, et al. The association of gastric leptin with oesophageal inflammation and metaplasia. *Gut*. 2008;57(1):16–24.
56. Kendall BJ, Macdonald GA, Hayward NK, et al. Leptin and the risk of Barrett's oesophagus. *Gut*. 2008;57(4):448–454.
57. Mohammed I, Cherkas LF, Riley SA, Spector TD, Trudgill NJ. Genetic influences in gastro-oesophageal reflux disease: a twin study. *Gut*. 2003;52(8):1085–1089.
58. Cameron AJ, Lagergren J, Henriksson C, Nyren O, Locke GR, Pedersen NL. Gastroesophageal reflux disease in monozygotic and dizygotic twins. *Gastroenterology*. 2002;122(1):55–59.
59. Dong J, Levine DM, Buas MF, et al. Interactions between genetic variants and environmental factors affect risk of esophageal adenocarcinoma and Barrett's esophagus. *Clin Gastroenterol Hepatol*. 2018;16(10):1598–1606. e1594.
60. Zhai R, Chen F, Liu G, et al. Interactions among genetic variants in apoptosis pathway genes, reflux symptoms, body mass index, and smoking indicate two distinct etiologic patterns of esophageal adenocarcinoma. *J Clin Oncol*. 2010;28(14):2445–2451.
61. Zhai R, Zhao Y, Liu G, et al. Interactions between environmental factors and polymorphisms in angiogenesis pathway genes in esophageal adenocarcinoma risk: a case-only study. *Cancer*. 2012;118(3):804–811.
62. Gyawali CP, Kahrilas PJ, Savarino E, et al. Modern diagnosis of GERD: the Lyon consensus. *Gut*. 2018;67(7):1351–1362.
63. Sifrim D, Zerbib F. Diagnosis and management of patients with reflux symptoms refractory to proton pump inhibitors. *Gut*. 2012;61(9):1340–1354.
64. Dent J, Vakil N, Jones R, et al. Accuracy of the diagnosis of GORD by questionnaire, physicians and a trial of proton pump inhibitor treatment: the diamond study. *Gut*. 2010;59(6):714–721.
65. Akdamar K, Ertan A, Agrawal NM, McMahon FG, Ryan J. Upper gastrointestinal endoscopy in normal asymptomatic volunteers. *Gastrointest Endosc*. 1986;32(2):78–80.
66. Savarino E, Zentilin P, Savarino V. NERD: an umbrella term including heterogeneous subpopulations. *Nat Rev Gastroenterol Hepatol*. 2013;10(6):371–380.
67. Parikh ND, Viana AV, Shah S, Laine L. Image-enhanced endoscopy is specific for the diagnosis of non-erosive gastroesophageal reflux disease. *Scand J Gastroenterol*. 2018;53(3):260–264.

68. Lv J, Liu D, Ma SY, Zhang J. Investigation of relationships among gastroesophageal reflux disease subtypes using narrow band imaging magnifying endoscopy. *World J Gastroenterol.* 2013;19(45):8391–8397.
69. Sharma P, Wani S, Bansal A, et al. A feasibility trial of narrow band imaging endoscopy in patients with gastroesophageal reflux disease. *Gastroenterology.* 2007;133 (2):454–464.
70. Chu CL, Zhen YB, Lv GP, et al. Microalterations of esophagus in patients with nonerosive reflux disease: in vivo diagnosis by confocal laser endomicroscopy and its relationship with gastroesophageal reflux. *Am J Gastroenterol.* 2012;107(6):864–874.
71. Lei WY, Vaezi MF, Naik RD, Chen CL. Mucosal impedance testing: a new diagnostic testing in gastroesophageal reflux disease. *J Formos Med Assoc.* 2020;119(11): 1575–1580.
72. Patel DA, Higginbotham T, Slaughter JC, et al. Development and validation of a mucosal impedance contour analysis system to distinguish esophageal disorders. *Gastroenterology.* 2019;156(6):1617–1626. e1611.
73. Ates F, Yuksel ES, Higginbotham T, et al. Mucosal impedance discriminates GERD from non-GERD conditions. *Gastroenterology.* 2015;148(2):334–343.
74. Saritas Yuksel E, Higginbotham T, Slaughter JC, et al. Use of direct, endoscopic-guided measurements of mucosal impedance in diagnosis of gastroesophageal reflux disease. *Clin Gastroenterol Hepatol.* 2012;10(10):1110–1116.
75. Shimamura Y, Inoue H, Rodriguez de Santiago E, et al. Characterization of intragastric pressure waveform in endoscopic pressure study integrated system: novel diagnostic device for gastroesophageal reflux disease. *Dig Endosc.* 2021;33(5):780–787.
76. Inoue H, Shimamura Y, Rodriguez de Santiago E, et al. Diagnostic performance of the endoscopic pressure study integrated system (EPSIS): a novel diagnostic tool for gastroesophageal reflux disease. *Endoscopy.* 2019;51(8):759–762.
77. Falk M, Van der Wall H, Falk GL. Differences between scintigraphic reflux studies in gastrointestinal reflux disease and laryngopharyngeal reflux disease and correlation with symptoms. *Nucl Med Commun.* 2015;36(6):625–630.
78. Park JS, Burton L, Van der Wall H, Falk GL. Modified reflux scintigraphy detects pulmonary microaspiration in severe gastro-esophageal and laryngopharyngeal reflux disease. *Lung.* 2021;199(2):139–145.
79. Burton L, Falk GL, Baumgart K, Beattie J, Simpson S, Van der Wall H. Esophageal clearance in laryngopharyngeal reflux disease: correlation of reflux scintigraphy and 24-hour impedance/pH in a cohort of refractory symptomatic patients. *Mol Imaging Radionucl Ther.* 2020;29(1):7–16.
80. Matsumura T, Arai M, Suzuki T, et al. Clinical utility of salivary pepsin measurement in patients with proton pump inhibitor-refractory gastroesophageal reflux disease symptoms: a prospective comparative study. *Esophagus.* 2020;17(3):339–347.
81. Li YW, Sifrim D, Xie C, Chen M, Xiao YL. Relationship between salivary pepsin concentration and esophageal mucosal integrity in patients with gastroesophageal reflux disease. *J Neurogastroenterol Motil.* 2017;23(4):517–525.
82. Haddad HA, He Z, Shaffer SE, Molle-Rios ZL. Salivary pepsin A detection related to gastro-oesophageal reflux episodes in children undergoing impedance probe monitoring. *Acta Paediatr.* 2020;109(11):2374–2379.
83. Woodland P, Singendonk MMJ, Ooi J, et al. Measurement of salivary pepsin to detect gastroesophageal reflux disease is not ready for clinical application. *Clin Gastroenterol Hepatol.* 2019;17(3):563–565.
84. Race C, Chowdry J, Russell JM, Corfe BM, Riley SA. Studies of salivary pepsin in patients with gastro-oesophageal reflux disease. *Aliment Pharmacol Ther.* 2019;49 (9):1173–1180.

85. Sethi S, Richter JE. Diet and gastroesophageal reflux disease: role in pathogenesis and management. *Curr Opin Gastroenterol*. 2017;33(2):107–111.
86. Mone I, Kraja B, Bregu A, et al. Adherence to a predominantly Mediterranean diet decreases the risk of gastroesophageal reflux disease: a cross-sectional study in a south eastern European population. *Dis Esophagus*. 2016;29(7):794–800.
87. Ness-Jensen E, Hveem K, El-Serag H, Lagergren J. Lifestyle intervention in gastroesophageal reflux disease. *Clin Gastroenterol Hepatol*. 2016;14(2):175–182. e171–173.
88. Freedberg DE, Kim LS, Yang YX. The risks and benefits of long-term use of proton pump inhibitors: expert review and best practice advice from the American gastroenterological association. *Gastroenterology*. 2017;152(4):706–715.
89. Shim YK, Kim N. The effect of H. *Korean J Gastroenterol*. 2017;70(1):4–12.
90. Sugano K. Vonoprazan fumarate, a novel potassium-competitive acid blocker, in the management of gastroesophageal reflux disease: safety and clinical evidence to date. *Ther Adv Gastroenterol*. 2018;11. 1756283X17745776.
91. Assakran BS, Alrakbi K, Alharbi MA, et al. Prevalence of asymptomatic hiatal hernia in obese patients during preoperative upper gastrointestinal endoscopy assessments and correlation with body mass index. *Cureus*. 2021;13(2):e13396.
92. Sloan S, Kahrilas PJ. Impairment of esophageal emptying with hiatal hernia. *Gastroenterology*. 1991;100(3):596–605.
93. Emerenziani S, Habib FI, Ribolsi M, et al. Effect of hiatal hernia on proximal oesophageal acid clearance in gastro-oesophageal reflux disease patients. *Aliment Pharmacol Ther*. 2006;23(6):751–757.
94. Kasapidis P, Vassilakis JS, Tzovaras G, Chrysos E, Xynos E. Effect of hiatal hernia on esophageal manometry and pH-metry in gastroesophageal reflux disease. *Dig Dis Sci*. 1995;40(12):2724–2730.
95. Pauwels A, Boecxstaens V, Andrews CN, et al. How to select patients for antireflux surgery? The ICARUS guidelines (international consensus regarding preoperative examinations and clinical characteristics assessment to select adult patients for antireflux surgery). *Gut*. 2019;68(11):1928–1941.
96. Jones MP, Sloan SS, Rabine JC, Ebert CC, Huang CF, Kahrilas PJ. Hiatal hernia size is the dominant determinant of esophagitis presence and severity in gastroesophageal reflux disease. *Am J Gastroenterol*. 2001;96(6):1711–1717.
97. Frazzoni M, De Micheli E, Grisendi A, Savarino V. Hiatal hernia is the key factor determining the lansoprazole dosage required for effective intra-oesophageal acid suppression. *Aliment Pharmacol Ther*. 2002;16(5):881–886.
98. Blouin RA, Warren GW. Pharmacokinetic considerations in obesity. *J Pharm Sci*. 1999;88(1):1–7.
99. Jacobson BC. Body mass index and the efficacy of acid-mediating agents for GERD. *Dig Dis Sci*. 2008;53(9):2313–2317.
100. Barak N, Ehrenpreis ED, Harrison JR, Sitrin MD. Gastro-oesophageal reflux disease in obesity: pathophysiological and therapeutic considerations. *Obes Rev*. 2002;3(1):9–15.
101. Sharma P, Vakil N, Monyak JT, Silberg DG. Obesity does not affect treatment outcomes with proton pump inhibitors. *J Clin Gastroenterol*. 2013;47(8):672–677.
102. Kim TJ, Lee H, Baek SY, et al. Metabolically healthy obesity and the risk of erosive esophagitis: a cohort study. *Clin Transl Gastroenterol*. 2019;10(9):e00077.
103. D'Silva M, Bhasker AG, Kantharia NS, Lakdawala M. High-percentage pathological findings in obese patients suggest that esophago-gastro-duodenoscopy should be made mandatory prior to bariatric surgery. *Obes Surg*. 2018;28(9):2753–2759.
104. Kristo I, Paireder M, Jomrich G, et al. Silent gastroesophageal reflux disease in patients with morbid obesity prior to primary metabolic surgery. *Obes Surg*. 2020; 30(12):4885–4891.

105. Ortiz V, Alvarez-Sotomayor D, Sáez-González E, et al. Decreased esophageal sensitivity to acid in morbidly obese patients: a cause for concern? *Gut Liver.* 2017; 11(3):358–362.
106. Kamboj AK, Katzka DA, Iyer PG. Endoscopic screening for Barrett's esophagus and esophageal adenocarcinoma: rationale, candidates, and challenges. *Gastrointest Endosc Clin N Am.* 2021;31(1):27–41.
107. Di Lorenzo N, Antoniou SA, Batterham RL, et al. Clinical practice guidelines of the European association for endoscopic surgery (EAES) on bariatric surgery: update 2020 endorsed by IFSO-EC, EASO and ESPCOP. *Surg Endosc.* 2020;34(6): 2332–2358.
108. Ortega J, Escudero MD, Mora F, et al. Outcome of esophageal function and 24-hour esophageal pH monitoring after vertical banded gastroplasty and Roux-en-Y gastric bypass. *Obes Surg.* 2004;14(8):1086–1094.
109. Jones KB. Roux-en-Y gastric bypass: an effective antireflux procedure in the less than morbidly obese. *Obes Surg.* 1998;8(1):35–38.
110. Tristão LS, Tustumi F, Tavares G, Bernardo WM. Fundoplication vs oral proton pump inhibitors for gastroesophageal reflux disease: a systematic review and *meta*-analysis of randomized clinical trials. *Esophagus.* 2021;18(2):173–180.
111. Nabi Z, Reddy DN. Update on endoscopic approaches for the management of gastroesophageal reflux disease. *Gastroenterol Hepatol (N Y).* 2019;15(7):369–376.
112. Fass R, Cahn F, Scotti DJ, Gregory DA. Systematic review and *meta*-analysis of controlled and prospective cohort efficacy studies of endoscopic radiofrequency for treatment of gastroesophageal reflux disease. *Surg Endosc.* 2017;31(12):4865–4882.
113. Perry KA, Banerjee A, Melvin WS. Radiofrequency energy delivery to the lower esophageal sphincter reduces esophageal acid exposure and improves GERD symptoms: a systematic review and *meta*-analysis. *Surg Laparosc Endosc Percutan Tech.* 2012;22(4):283–288.

CHAPTER 4

Impact of obesity on Barrett's esophagus and esophageal adenocarcinoma

Nasim Parsa and Allon Kahn
Division of Gastroenterology and Hepatology, Mayo Clinic College of Medicine & Science, Scottsdale, AZ, United States

Introduction

The incidence of esophageal adenocarcinoma (EAC) in western countries has increased more than sixfold in the last few decades.[1] Similarly, there has been a substantial increase in obesity prevalence, with approximately 36% of adults in the United States currently considered to be obese.[2] Barrett's esophagus (BE), which is a columnar metaplastic change in the distal esophagus, is the only known precursor for EAC.[3] Previous studies have shown a strong association between obesity and increased risk of BE and EAC. Stein et al. evaluated the association between the body mass index (BMI) and the presence of BE and reported a significantly higher BE rate in obese patients, with a reported 10% increase in BE risk per each 10 pounds increase in total body weight.[4] Abdallah and colleagues studied the relationship between the length of BE mucosa and BMI and reported a significantly higher BMI among patients with long-segment BE compared with patients with short-segment BE (mean BMI 32.7 vs 30.3, $p = 0.001$).[5] In an international pooled analysis including approximately 4000 patients with EAC and 10,000 controls, there was a significant dose−response correlation between the BMI and EAC, with the greatest risk observed among patients with the most severe obesity (OR 4.7, 2.7, and 2.3, for obesity class III, II, and I, respectively).[6] In this context, several studies have shown a significant correlation between central obesity and an increased risk of BE, dysplastic progression, and EAC. Corley et al. demonstrated a significant correlation between BE risk and abdominal circumference greater than 80 cm compared with less than 80 cm (OR 2.2, 95% CI: 1.2−4.1).[7] In a similar study, an abdominal

Obesity and Esophageal Disorders
DOI: https://doi.org/10.1016/B978-0-323-98365-5.00001-2

diameter index ≥0.6 was found to be a significant independent predictor of BE (OR 5.7, 95% CI: 1.2–2.5), and a stronger BE predictor compared with the BMI itself.[8] In a systematic review and *meta*-analysis including 40 studies, abdominal obesity was significantly associated with the risk of BE development (OR 1.9, 95% CI: 1.5–2.5).[9] In a prospective cohort study of obesity and risk of EAC, waist–hip ratio, a measure of central obesity was associated with increased EAC risk; even in people with normal BMI.[10] In a systematic review and *meta*-analysis including 7 studies and 913,182 patients, both higher waist circumference and waist–hip ratio were significantly associated with higher EAC risk (RR 2.0, 95% CI: 1.3–3.2 and RR 1.9, 95% CI: 1.0–3.7, respectively).[11]

Two pathways have been associated with the BE and EAC risk and central obesity. The first is a gastroesophageal reflux disease (GERD)-dependent mechanism. Central obesity causes increased intraabdominal pressure and increases the risk of a hiatal hernia, which promotes the lower esophageal sphincter relaxations that can result in retrograde movement of gastric contents into the esophagus.[12–14] The second mechanism is the GERD-independent pathway, which is mediated by systemic inflammation through hyperleptinemia, insulin resistance, and an altered esophageal microbiome. In a systematic review and *meta*-analysis by Singh et al., patients with central obesity were found to have an increased risk for BE, irrespective of their BMI. Moreover, the significant association between BE and central obesity still persisted even after adjusting for symptomatic reflux (OR 2.0, 95% CI: 1.4–2.9), indicating the importance of GERD-independent mechanisms in central obesity and the development of BE.[9]

Mechanisms of carcinogenesis

Local inflammation from acid and bile reflux

The chronic exposure of esophageal squamous epithelium to gastric acid promotes a local inflammatory process that results in the development of columnar metaplasia and progression to dysplasia in BE.[15] Additionally, bile reflux has been associated with an increased risk of BE through local inflammatory mediators, with studies demonstrating genomic instability and DNA damage in esophageal squamous cells exposed to bile.[16,17] Moreover, studies have shown that both bile and acid can promote genomic instability and DNA damage in BE, regardless of the current dysplasia grade.[18,19]

Systemic inflammation

The visceral adipose tissue (omentum, mesentery, epiploic, epicardial, gonadal, and retroperitoneal fat) contains an increased number of proinflammatory cells that produce proinflammatory cytokines such as tumor necrosis factor-α (TNF- α), interleukin (IL)-1b, and IL-6, which contribute to a systemic inflammatory state and carcinogenesis.[20,21] In a prospective case–control study on the role of inflammatory markers in EAC, soluble TNF receptor 2 was significantly associated with EAC (OR 2.6, 95% CI: 1.5–4.6), accounting for 33% of the effect of waist circumference on EAC risk. Additional markers close to the adjusted significance threshold included C-reactive protein (CRP), serum amyloid A, lipocalin-2, resistin, IL-3, IL-17A, soluble IL-6 receptor, and soluble vascular endothelial growth factor receptor 3.[22] The specific cell signaling mechanisms underlying the relationship between adipocyte pro-inflammatory markers and dysplasia/ EAC risk include TNF-alpha-mediated c-myc oncogene induction and IL-6 activation of the STAT3 antiapoptosis pathway.[23,24] Finally, in a prospective study evaluating the predictors of EAC in patients with BE, CRP and IL-6 levels above the median were associated with an 80% and almost twofold increased risk of progression to EAC, respectively.[20]

Adipokines

Leptin and adiponectin are adipokines that are secreted by adipocytes. Leptin release activates the release of the proinflammatory cytokines.[25] It has been shown that leptin receptors are present on the esophageal epithelium, with studies demonstrating high levels of leptin receptors on BE and EAC cells, resulting in increased cell proliferation and reduced apoptosis.[26–28] Moreover, leptin receptors are similar to the receptors of some proinflammatory cytokines such as TNF-α, IL-2, and IL-6, which contribute to the obesity-related inflammatory state.[29] In a case–control study, a significant positive association was noted between the circulatory levels of leptin and BE, irrespective of the obesity, GERD, and insulin levels (OR 3.2, 95% CI: 1.2–8.1).[30] Elevated leptin levels have been associated with twofold and eightfold higher risk of BE.[31,32] Another study showed a significant direct association between central obesity and advanced tumor and nodal stage in EAC, and upregulated leptin receptors.[28]

In contrast, adiponectin is an antiinflammatory hormone that acts in counterbalance to leptin. It regulates cell proliferation, induces apoptosis, inhibits growth factors, and increases insulin sensitivity, with circulatory levels inversely proportional to the amount of body adipose tissue.[27,33] Similar to leptin, adiponectin receptors are also expressed on BE and EAC cells. In a study of esophagectomy specimens of patients with EAC, significantly higher levels of adiponectin receptors were observed in specimens with less advanced tumor stage.[26] Lower serum adiponectin levels have also been documented in BE, when compared with control patients.[34] Finally, adiponectin receptor levels exhibit a nonlinear inverse association with the risk of BE progression to EAC.[35]

Insulin and insulin-like growth factor

Increased body fat mass and circulatory free fatty acids promote peripheral insulin resistance that leads to an increase in the levels of circulatory insulin.[36] Leptin and adiponectin have opposite effects on insulin and insulin resistance, with elevated serum leptin being strongly associated with insulin resistance, while high adiponectin increases peripheral tissue insulin sensitivity.[37,38] In a large population-based case–control study, diabetes was associated with a nearly 50% increase in the risk of BE, independent of other known risk factors such as GERD and BMI.[39] Hyperinsulinemia and insulin resistance promote cell mitogenesis, prevent apoptosis, and alter the insulin-like growth factor (IGF) cascade, resulting in an increase in free active IGF levels.[40] Overexpression of IGF-1 receptors has also been shown to be strongly associated with the progression of BE to EAC. In contrast, blockade of IGF-1 receptors leads to apoptosis of EAC cells.[41] In a study on the role of central obesity and the IGF-1 in EAC and esophageal squamous-cell carcinoma, serum IGF-1 levels were highest in EAC patients ($p < 0.01$) and higher in patients with central obesity versus nonobese patients ($p < 0.05$).[42] A study of resected EAC specimens showed that the IGF-1 receptor expression was significantly higher in patients with central obesity compared with patients with normal weight ($p = 0.02$). Further, the same study demonstrated a higher disease-specific survival rate in patients with negative IGF-1 receptor expression compared with increased tumor IGF-1 receptor expression (60 vs 23 months, $p = 0.02$).[43] Finally, in a study on inherited genetic variation in the IGF pathway and BE/EAC risk, the germline variation in the IGF pathway was reported to influence the risk of BE and EAC.[44]

Diet and gut microbiome

Several studies have shown the impact of diet and gut microbiome in the regulation of inflammation and metabolism. Chen et al. reported a higher rate of BE in animals fed with a high-fat dairy diet, compared with those with a low-fat diet and proposed that an increase in taurine conjugates in the bile following fat consumption may promote the development of BE.[45] In a population study using a food frequency questionnaire to determine the association between dietary antioxidants and risk of BE and EAC, an inverse association was reported between high dietary β-carotene (OR 0.4, 95% CI: 0.2–1.0) and vitamin E and antioxidants (OR 0.4, 95% CI: 0.2–0.6) and the risk of dysplastic BE and EAC.[46] In a *meta*-analysis including 16,885 patients from 15 studies, there was a significant association between dietary fiber intake and reduced risk for BE and EAC, with a pooled odds ratio of 0.5 (95% CI: 0.4–06) for the highest compared with the lowest dietary fiber intake.[47] GERD and obesity may alter the esophageal microbiota, with data suggesting increased numbers of gram-negative anaerobes and microaerophiles of the phyla Bacteroides, Proteobacteria, and Fusobacteria in BE patients.[48,49] Kaakoush and colleagues examined the effect of high-fat diet on the esophageal microbiota in rats and reported that high-fat diet can significantly alter the esophageal microbiota and host esophageal gene expression. In particular, Fusobacterium, Rothia, and Granulicatella showed consistent correlations across a range of metabolic and gene markers.[50] Similarly, mice fed a high-fat diet developed esophageal dysplasia and tumors more rapidly than mice fed the control diet, which was associated with a shift in the gut microbiota and an increased ratio of neutrophils to natural killer cells in esophageal tissues.[51] Finally, in a report by the World Cancer Research Fund including 46 studies and 31,000 EAC cases, there was a strong association between obesity and the risk of EAC, with some data suggesting a decreased risk of EAC in individuals who consume vegetables and are physically active.[52]

Bariatric surgery and the risk of Barrett's esophagus and esophageal adenocarcinoma

As the prevalence of obesity continues to rise, so too has the demand for bariatric surgery to address it. The number of bariatric surgeries performed annually increased from 8631 in 1993 to 1,62,969 in 2016 and the trend upward continues.[53] These surgical procedures are designed and utilized for the principal indication of weight loss, which has clear beneficial implications for the risk of developing BE and EAC, as outlined earlier. However, they each significantly modify the upper gastrointestinal anatomy and physiology with respect to GERD, which remains one of the most important risk factors for developing BE and EAC.

According to a recent population-based US study of over 1,14,000 patients undergoing bariatric surgery from 2008 to 2014, there have been striking trends in the type of surgery being performed nationally.[54] While laparoscopic adjustable gastric banding was previously favored, its use dropped precipitously beginning in 2010 and by 2014 it represented only 32% of outpatient and 1% of inpatient surgeries. In contrast, vertical sleeve gastrectomy (VSG) has experienced a dramatic increase in utilization and is now the most utilized bariatric surgical technique, representing 54% of cases in 2014. Roux-en-Y gastric bypass (RYGB) has seen a corresponding decrease in utilization, but to a vastly lesser extent than banding.

It is worth considering what implications the meteoric rise in VSG popularity may have on the risk of BE and EAC, particularly in contrast to RYGB. Initial studies assessing the effect of VSG on GERD were conflicting. A systematic review and *meta*-analysis of 15 studies representing 2163 patients demonstrated that GERD prevalence was reduced in 4 studies and increased in 7.[55] However, these studies relied almost entirely on self-reported symptom questionnaires and not ambulatory pH monitoring, endoscopic esophagitis, or other objective findings. It was postulated that VSG may increase GERD by increasing intragastric pressure, but that this effect may be counteracted in some patients by weight loss. Subsequent studies have clearly demonstrated the reflexogenic nature of the VSG, due to a multitude of mechanisms, including loss of the angle of His, hypotensive LES pressure, lack of gastric compliance, and relative gastric stasis.[56]

The lack of correlation between GERD symptoms and objective GERD findings in VSG is particularly worrisome in relation to BE and EAC risk. In a study of esophageal manometry in 53 patients following VSG, increased intragastric pressure occurred in 77% and was not symptomatic, despite a concomitant increase in impedance reflux events.[57] To investigate this relationship further, a recent *meta*-analysis of 10 studies and 680 patients evaluated the prevalence of BE in patients undergoing upper endoscopy after VSG.[58] The pooled prevalence of BE was 11.6% and this was not associated with the rate of symptomatic GERD. While routine screening upper endoscopy before all bariatric surgeries is not mandated by existing guidelines, the American Society for Gastrointestinal Endoscopy produced a joint statement in 2015 with the Society of Gastrointestinal and Endoscopic Surgeons and American Society for Metabolic and Bariatric Surgery and stated that erosive esophagitis and the presence of a hiatal hernia were a relative contraindication to performing VSG due to the risk of postoperative GERD.[59] In light of the high prevalence of postoperative GERD following VSG, a screening endoscopy is warranted in this population and the surgery should be reconsidered in the presence of symptomatic GERD.

In contrast, RYGB has been long considered an effective antireflux surgery in addition to its prominent role in weight loss for patients with obesity. RYGB has been used as a salvage method for failed fundoplication, where it has shown excellent efficacy, reducing reflux symptoms in over 80% of patients.[60] Therefore it is notable that reflux occurring after VSG may be best handled by conversion to RYGB. This has been shown in a recent study of 10 patients with symptomatic GERD and BE following VSG who were treated effectively in this manner.[61]

In summary, for patients with medically complicated obesity undergoing evaluation for bariatric surgery, GERD and risk of BE and EAC must be considered in preoperative evaluation, surgical procedure selection, and postoperative monitoring.

Conclusion

The incidence of both obesity and EAC have significantly increased in the last decades. Obesity, specifically central obesity is a risk factor for BE and EAC through increased systemic inflammation, insulin resistance, IGF-1, and immune cell alterations. In addition, obesity and a high-fat diet modulate the gut microbiota with an increase in bacterial species known to be associated with gastrointestinal carcinogenesis. At the same time, bariatric surgeries such as VSG can cause worsening GERD and progression to BE. Further studies are needed to evaluate the role of diet, lifestyle, and bariatric interventions in modulating the risk of BE and EAC in obese patients.

Disclosures

Dr. Nasim Parsa disclosed no financial relationships.
Dr. Allon Kahn has received research support from NinePoint Medical.

References

1. Hur C, Miller M, Kong CY, et al. Trends in esophageal adenocarcinoma incidence and mortality. *Cancer*. 2013;119:1149–1158.
2. Ogden CL, Carroll MD, Fryar CD, et al. Prevalence of obesity among adults and youth: United States, 2011–2014. *NCHS Data Brief*. 2015;219:1–8.
3. Rustgi AK, El-Serag HB. Esophageal carcinoma. *N Engl J Med*. 2014;371:2499–2509.
4. Stein DJ, El-Serag HB, Kuczynski J, et al. The association of body mass index with Barrett's esophagus. *Aliment Pharmacol Ther*. 2005;22:1005–1010.
5. Abdallah J, Maradey-Romero C, Lewis S, et al. The relationship between length of Barrett's esophagus mucosa and body mass index. *Aliment Phamarcol Ther*. 2015; 41:137–144.
6. Hoyo C, Cook MB, Kamangar F, et al. Body mass index in relation to oesophageal and oesophagogastric junction adenocarcinomas: a pooled analysis from the International BEACON Consortium. *Int J Epidemiol*. 2012;41:1706–1818.
7. Corley DA, Kubo A, Levin TR, et al. Abdominal obesity and body mass index as risk factors for Barrett's esophagus. *Gastroenterology*. 2007;133:34–41.
8. Baik D, Sheng J, Schlaffer K, et al. Abdominal diameter index is a strong predictor of prevalent Barretts esophagus than BMI or waist-to-hip ratio. *Dis Esophagus*. 2017; 30:1–6.
9. Singh S, Sharma AN, Murad MH, et al. Central adiposity is associated with increased risk of esophageal inflammation, metaplasia, and adenocarcinoma: a systematic review and *meta*-analysis. *Clin Gastroenterol Hepatol*. 2013;11:1399–1412.
10. O'Doherty MG, Freedman ND, Hollenbeck AR, et al. A prospective cohort study of obesity and risk of oesophageal and gastric adenocarcinoma in the NIH-AARP diet and health study. *Gut*. 2012;61:1261–1268.
11. Du X, Hidayat K, Shi BM. Abdominal obesity and gastroesophageal cancer risk: systematic review and *meta*-analysis of prospective studies. *Biosci Rep*. 2017;37.
12. Schneider JH, Keuper M, Keonigsrainer A, et al. Transient lower esophageal sphincter relaxation in morbid obesity. *Obes Surg*. 2009;19:595–600.
13. Del Grande LDM, Heberlla FAM, Katayama RC, et al. Transdiaphragmatic pressure gradient (TPG) has a central role in the pathophysiology of gastroesophageal reflux disease (GERD) in the obese and it correlates with abdominal circumference but not with body mass index (BMI). *Obes Surg*. 2020;30:1424–1428.
14. Che F, Nguyen B, Cohen A, et al. Prevalence of hiatal hernia in the morbidity obese. *Surg Obes Relat Dis*. 2013;9:920–924.
15. Fitzgerald RC. Barrett's oesophagus and oesophageal adenocarcinoma: how does acid interfere with cell proliferation and differentiation? *Gut*. 2005;54(suppl 1): i21–i26.
16. O'Sullivan KE, Phelan JJ, O'Hanlon C, et al. The role of inflammation in cancer of the esophagus. *Expert Rev Gastroenterol Hepatol*. 2014;8:749–760.
17. McQuaid KR, Laine L, Fennerty MB, et al. Systematic review: the role of bile acids in the pathogenesis of gastro-oesophageal reflux disease and related neoplasia. *Aliment Pharmacol Ther*. 2011;34:146–165.
18. Weaver JMJ, Ross-Innes CS, Shannon N, et al. Ordering of mutations in preinvasive disease stages of esophageal carcinogenesis. *Nat Genet*. 2014;46:837–843.
19. Khara HS, Jackson SA, Nair S, et al. Assessment of mutational load in biopsy tissue provides additional information about genomic instability to histological classifications of Barrett's esophagus. *J Gastrointest Cancer*. 2014;45:137–145.
20. Hardikar S, Onstad L, Song X, et al. Inflammation and oxidative stress markers and esophageal adenocarcinoma incidence in a Barrett's esophagus cohort. *Cancer Epidemiol Biomarkers Prev*. 2014;23:2393–2403.

21. Lysaght J, van der Stok EP, et al. Pro-inflammatory and tumour proliferative properties of excess visceral adipose tissue. *Cancer Lett.* 2011;312:62–72.
22. Cook MB, Barnett MJ, Bock CH, et al. Pre diagnostic circulating markers of inflammation and risk of oesophageal adenocarcinoma: a study with the national cancer institute cohort consortium. *Gut.* 2019;68:960–968.
23. Tselepis C, Perry I, Dawson C, et al. Tumour necrosis factor-alpha in Barrett's oesophagus: a potential novel mechanism of action. *Oncogene.* 2002;21:6071–6081.
24. Dvorak K, Chavarria M, Payne CM, et al. Activation of the interleukin-6/STAT3 antiapoptotic pathway in esophageal cells by bile acids and low pH: relevance to Barrett's esophagus. *Clin Cancer Res.* 2007;13:5305–5313.
25. Ogunwobi O, Mutungi G, Beales IL. Leptin stimulates proliferation and inhibits apoptosis in Barrett's esophageal adenocarcinoma cells by cyclooxygenase-2-dependent, prostaglandin-E2-mediated transactivation of the epidermal growth factor receptor and c-Jun NH2-terminal kinase activation. *Endocrinology.* 2006;147:4505–4516.
26. Howard JM, Cathcart MC, Healy L, et al. Leptin and adiponectin receptor expression in oesophageal cancer. *Br J Surg.* 2014;101:643–652.
27. Beales IL, Garcia-Morales C, Ogunwobi OO, et al. Adiponectin inhibits leptin-induced oncogenic signalling in oesophageal cancer cells by activation of PTP1B. *Mol Cell Endocrinol.* 2014;382:150–158.
28. Howard JM, Beddy P, Ennis D, et al. Associations between leptin and adiponectin receptor upregulation, visceral obesity, and tumour stage in oesophageal and junctional adenocarcinoma. *Br J Surg.* 2010;97:1020–1027.
29. Fernandez-Riejos P, Najib S, Santos-Alvarez J, et al. Role of leptin in the activation of immune cells. *Mediators Inflamm.* 2010;2010(1–8):568343.
30. Rubenstein JH, Morgenstern H, Mcconell D, et al. Associations of diabetes mellitus, insulin, leptin and ghrelin with gastroesophageal reflux and Barrett's esophagus. *Gastroenterology.* 2013;145:1237–1244.
31. Kendall BJ, Macdonald GA, Hayward NK, et al. Study of digestive health. Leptin and the risk of Barrett's oesophagus. *Gut.* 2008;57:448–454.
32. Garcia JM, Splenser AE, Kramer J, et al. Circulating inflammatory cytokines and adipokines are associated with increased risk of Barrett's esophagus: a case control study. *Clin Gastroenterol Hepatol.* 2014;12:229–238.
33. Barb D, Williams CJ, Neuwirth AK, et al. Adiponectin in relation to malignancies: a review of existing basic research and clinical evidence. *Am J Clin Nutr.* 2007;86: 858–866.
34. Mokrowiecka A, Daniel P, Jasinska A, et al. Serum adiponectin, resistin, leptin concentration and central adiposity parameters in Barrett's esophagus patients with and without intestinal metaplasia in comparison to healthy controls and patients with GERD. *Hepatogastroenterology.* 2012;59:2395–2399.
35. Duggan C, Onstad L, Hardikar S, et al. Association between markers of obesity and progression from Barrett's esophagus to esophageal adenocarcinoma. *Clin Gastroenterol Hepatol.* 2013,11:934 943.
36. Sears B, Perry M. The role of fatty acids in insulin resistance. *Lipids Health Dis.* 2015;14:121. Available from: https://doi.org/10.1186/s12944-015-0123-1.
37. Yadav A, Kataria MA, Saini V, et al. Role of leptin and adiponectin in insulin resistance. *Clin Chim Acta.* 2013;417:80–84.
38. Kadowaki T, Yamauchi T, Kubota N, et al. Adiponectin and adiponectin receptors in insulin resistance, diabetes, and the metabolic syndrome. *J Clin Invest.* 2006; 116:1784–1792.
39. Iyer PG, Borah BJ, Heien HC, et al. Association of Barrett's esophagus with type II diabetes mellitus: results from a large population-based case-control study. *Clin Gastroenterol Hepatol.* 2013;11:1108–1114.

40. Rubin R, Baserga R. Insulin-like growth factor-I receptor. Its role in cell proliferation, apoptosis, and tumorigenicity. *Lab Invest.* 1995;73:311–331.
41. Iravani S, Zhang HQ, Yuan ZQ, et al. Modification of insulin-like growth factor 1 receptor, c-Src, and Bcl-XL protein expression during the progression of Barrett's neoplasia. *Hum Pathol.* 2003;34:975–982.
42. Doyle SL, Donohoe CL, Finn SP, et al. IGF-1 and its receptor in esophageal cancer: association with adenocarcinoma and visceral obesity. *Am J Gastroenterol.* 2012;107:196–204.
43. Donohoe CL, Doyle SL, McGarrigle S, et al. Role of the insulin-like growth factor 1 axis and visceral adiposity in oesophageal adenocarcinoma. *Br J Surg.* 2012; 99:387–396.
44. Dighe SG, Chen J, Yan L, et al. Germline variation in the insulin-like growth factor pathway and risk of Barrett's esophagus and esophageal adenocarcinoma. *Carcinogenesis.* 2021;17(42):369–377.
45. Chen KH, Mukaisho K, Sugihara H, et al. High animal-fat intake changes the bile-acid composition of bile juice and enhances the development of Barrett's esophagus and esophageal adenocarcinoma in a rat duodenal-contents reflux model. *Cancer Sci.* 2007;98:1683–1688.
46. Ibiebele TI, Hughes MC, Nagle CM, et al. Dietary antioxidants and risk of Barrett's esophagus and adenocarcinoma of the esophagus in an Australian population. *Int J Cancer.* 2013;133:214–224.
47. Sun L, Zhang Z, Xu J, et al. Dietary fiber intake reduces risk for Barrett's esophagus and esophageal cancer. *Crit Rev Food Sci Nutr.* 2017;57:2749–2757.
48. Gall A, Fero J, McCoy C, Claywell BC, et al. Bacterial composition of the human upper gastrointestinal tract microbiome is dynamic and associated with genomic instability in a Barrett's esophagus cohort. *PLoS One.* 2015;10:e0129055.
49. Snider EJ, Compres G, Freedberg DE, et al. Alterations to the esophageal microbiome associated with progression from Barrett's esophagus to esophageal adenocarcinoma. *Cancer Epidemiol Biomark Prev.* 2019;28:1687–1693.
50. Kaakoush NO, Lecomte V, Maloney CA, et al. Cross-talk among metabolic parameters, esophageal microbiota, and host gene expression following chronic exposure to an obesogenic diet. *Sci Rep.* 2017;7:45753.
51. Munch NS, Fang HY, Ingermann J, et al. High-fat diet accelerates carcinogenesis in a mouse model of Barrett's esophagus via interleukin 8 and alterations to the gut microbiome. *Gastroenterology.* 2019;157:492–506.
52. World Cancer Research Fund/American Institute for Cancer Research. Continuous update project expert report 2018. Diet, nutrition, physical activity, and oesophageal cancer. [cited 2021 August 23]. Available from dietandcancerrerport.org.
53. Campos GM, Khoraki J, Browning MG, et al. Changes in utilization of bariatric surgery in the United States from 1993 to 2016. *Ann Surg.* 2020;271:201–209.
54. Abraham A, Ikramuddin S, Jahansouz C, et al. Trends in bariatric surgery: procedure selection, revisional surgeries, and readmissions. *Obes Surg.* 2016;26:1371–1377.
55. Chiu S, Birch DW, Shi X, et al. Effect of sleeve gastrectomy on gastroesophageal reflux disease: a systematic review. *Surg Obes Relat Dis.* 2011;7:510–515.
56. Bou Daher H, Sharara AI. Gastroesophageal reflux disease, obesity and laparoscopic sleeve gastrectomy: the burning questions. *World J Gastroenterol.* 2019;25:4805–4813.
57. Mion F, Tolone S, Garros A, et al. High-resolution impedance manometry after sleeve gastrectomy: increased intragastric pressure and reflux are frequent events. *Obes Surg.* 2016;26:2449–2456.
58. Qumseya BJ, Qumsiyeh Y, Ponniah SA, et al. Barrett's esophagus after sleeve gastrectomy: a systematic review and *meta*-analysis. *Gastrointest Endosc.* 2021;93: 343–352. e2.

59. Foletto M, De Marchi F, Bernante P, et al. Late gastric pouch necrosis after lap-band, treated by an individualized conservative approach. *Obes Surg*. 2005;15:1487–1490.
60. Coakley KM, Groene SA, Colavita PD, et al. Roux-En-Y gastric bypass following failed fundoplication. *Surg Endosc*. 2018;32:3517–3524.
61. Felsenreich DM, Langer FB, Bichler C, et al. Roux-en-Y gastric bypass as a treatment for Barrett's esophagus after sleeve gastrectomy. *Obes Surg*. 2020;30:1273–1279.

CHAPTER 5

Obesity and esophageal dysmotility

Kevin Shah[1], Francesca Raffa[2] and Rishi D. Naik[2]
[1]Department of Internal Medicine, Vanderbilt University Medical Center, Nashville, TN, United States
[2]Division of Gastroenterology, Vanderbilt University Medical Center, Nashville, TN, United States

Introduction

Obesity is a global health problem that impacts greater than half a billion individuals.[1] Usually progressive, with multifactorial etiologies including excessive food intake, genetic predisposition, and absence of physical activity, obesity is a known risk factor for multiple systemic diseases.[2] Obesity continues to be a risk factor for gastroesophageal reflux disease (GERD) but has also shown a correlation with esophageal dysmotility. A recent study investigated manometric data in morbidly obese patients and showed a higher prevalence of abnormal manometric parameters including impaired lower esophageal sphincter (LES) relaxation and disorders of esophageal body peristalsis (such as distal esophageal spasm, ineffective esophageal motility (IEM), and hypercontractile disorders) as shown in Fig. 5.1.[3]

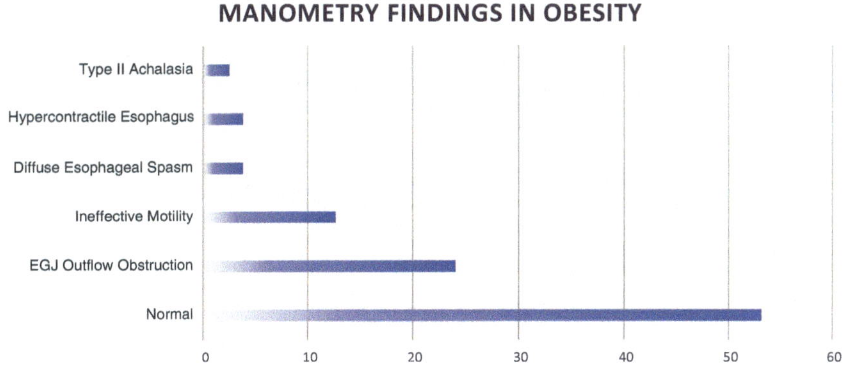

Figure 5.1 Manometric findings in obesity. *Data from Popescu AL, Costache RS, Costache DO, et al. Manometric changes of the esophagus in morbidly obese patients.* Exp Ther Med. *2021;21(6):604.*

Obesity and Esophageal Disorders
DOI: https://doi.org/10.1016/B978-0-323-98365-5.00004-8

In this chapter, we will discuss the pathophysiology, clinical presentation, and diagnosis of common esophageal motility disorders and outline the impact that obesity has on esophageal motility.

Impact of obesity on esophageal motility

The impact of obesity on esophageal motility is increasingly being recognized. Studies have shown a correlation with generalized hypomotility of the esophagus. In a 2014 prospective study of 53 obese patients, 51% were found to have abnormal manometry. Of those patients, dysmotility in the esophageal body was noted in 74%, while 11% had disorder of the esophagogastric junction.[4] Another study investigated manometric changes in 79 morbidly obese patients, which also showed that nearly half of the patients (46%) had abnormal manometry findings. Of these, 12.7% (10 patients) had IEM, 3.8% with distal esophageal spasm (DES), and 3.8% (3 patients) had jackhammer esophagus. Interestingly, the majority of motility disorders noted were found in asymptomatic patients. This could either represent a loss of visceral sensation in obese patients or could be incidental findings in a group being studied.[3] This finding further supported a prior 1999 study on asymptomatic motility disorders in obese patients, where of the 68 out of 111 obese patients with manometric abnormalities (DES, hypercontractile esophagus, nonspecific esophageal motility disorder), 70% did not have esophageal symptoms.[5]

In addition to esophageal body dysfunction, it is suspected that patients with obesity are prone to decreased LES pressures (Fig. 5.2). This might predispose obese patients to gastroesophageal reflux (GERD), which is a known obesity associated morbidity.[6] A 2009 study investigated dysfunction of the LES in 47 morbidly obese patients. Patients without GERD symptoms were divided into subgroups based on BMI (35–39.9, 40.0–49.9, and >50.0). These patients were compared to 15 normal individuals (BMI <30.0). Morbidly obese patients (BMI >40) were more often found to have a lower resting LES pressure. Correlations of decreased LES pressure and obesity have been described, generally in individuals with symptomatic GERD.[7] Our group has previously shown the impact of obesity in GERD with increased transient lower esophageal relaxations, increased prevalence of hiatal hernia, and decreased esophageal acid clearance (Fig. 5.3A), which leads to GERD and complications of GERD including esophagitis, Barrett's esophagus, and esophageal cancer (Fig. 5.3B).[8]

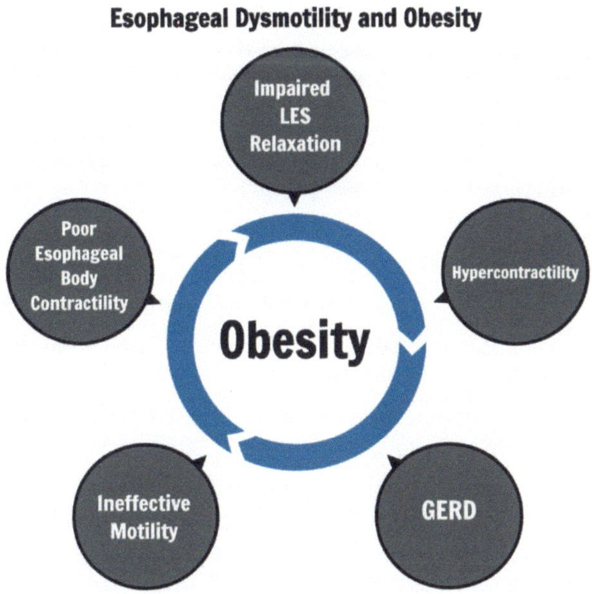

Figure 5.2 Esophageal dysmotility and obesity.

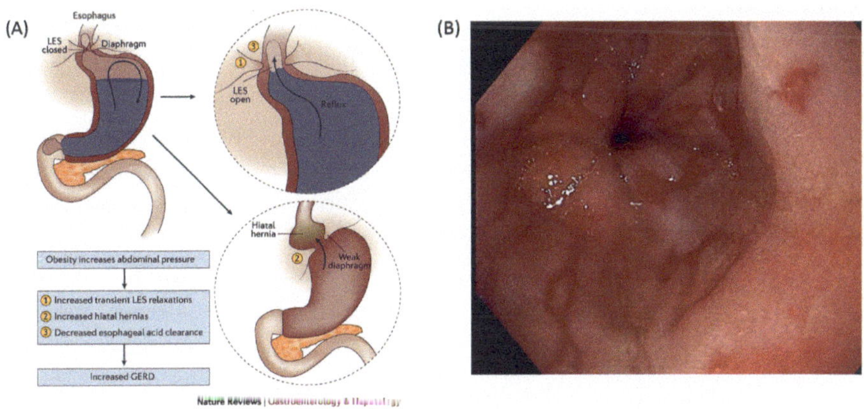

Figure 5.3 (A) Mechanisms of action for GERD in obesity, (B) Complications of GERD showing esophagitis on upper endoscopy. *From Naik RD, et al. Consequences of bariatric surgery on oesophageal function in health and disease.* Nat Rev Gastroenterol Hepatol. *2015. https://doi.org/10.1038/nrgastro.2015.202.*

A biochemical mechanism between obesity and esophageal physiology has been proposed. In a 2015 study, fasting plasma leptin concentrations were measured in nine patients without dysmotility and eight with dysmotility.[9] A study defined leptin score was calculated based on the ratio of the raw leptin value for the patient divided by the normal predicted value according to sex and BMI. Leptin scores were significantly higher in the group with esophageal dysmotility compared to without (99% vs 66%).[9] Given these findings, it has been suggested that elevated plasma leptin levels could drive a feedback mechanism for esophageal motility.

Overall, patients with obesity are more likely to have esophageal motility disorder, with many potentially being asymptomatic. These patients are also at risk for gastroesophageal reflux, likely from decreased LES tone. Although patients with morbid obesity may undergo esophageal testing when under consideration for restrictive bariatric surgery, they may benefit from motility testing regardless of surgical planning. Current evidence supports that this patient population is at higher risk for esophageal dysfunction. Additionally, the comorbidities seen with esophageal motility disorders (i.e., GERD or diabetes) are also prevalent in obese patients. This correlation is well known; however, additional investigation is needed to understand this relationship.

Evaluation

Once a mucosal or structural abnormality has been excluded with upper endoscopy, patients suspected of esophageal dysmotility should undergo esophageal function testing.[10] Esophageal high-resolution manometry (HRM), in combination with the Chicago Classification framework, is considered the gold standard test for diagnosing and classifying esophageal motility disorders.[11,12] Manometry can also help in the diagnosis of a hiatal hernia (HH), a common and surgically correctable finding in obese individuals, in which increased intraabdominal pressure and visceral adiposity affect LES pressure.[3,13,14] In a recent study, preoperative HRM showed a higher sensitivity (90.9%) in identifying HH in obese patients undergoing bariatric surgery, compared to barium swallow (45.5%) and upper gastrointestinal endoscopy (72.7%), assuming intraoperative diagnosis as a reference standard.[15]

HRM involves the transnasal placement of a catheter with multiple pressor sensors along the length of the entire esophagus.[16] After an acclimation period, basal pressure measurements are obtained and anatomic landmarks are identified. The patient then performs 10 single wet swallows during which pressure measurements are recorded, allowing assessment of pharyngeal, esophageal, and lower esophageal functions simultaneously. The 2021 Chicago Classification version 4.0 (CCv4.0) expanded the standardized manometry protocol to include both supine and upright positions as well as provocative testing to improve the diagnostic sensitivity and specificity of HRM.[12] Provocative testing recommended by CCv4.0 includes manometric evaluation with multiple rapid swallows and a rapid drink challenge. Pharmacologic provocation with amyl nitrite or cholecystokinin may also be performed if available, but this is restricted to only few esophageal physiology labs in the country.[12]

If manometric findings are equivocal and/or there is suspicion for an esophagogastric junction outflow obstruction (EGJOO), CCv4.0 recommends performing supplementary testing, either timed barium esophagram with a barium tablet or endoluminal functional lumen imaging planimetry (FLIP).[12] FLIP utilizes high-resolution impedance planimetry to obtain a three-dimensional image of the esophageal lumen at the time of endoscopy. A catheter with a distal overlying balloon is placed transorally into the esophagus, and once the anatomic point of interest is reached, the balloon is distended. Electrodes along the catheter capture impedance data as well as the intraballoon pressure, generating a cross-sectional area measurement. This provides a real-time assessment of esophageal distensibility.[17,18] FLIP can be particularly advantageous in confirming the diagnosis of suspected achalasia patients with relatively low integrated relaxation pressure (IRP) (precluding diagnosis of achalasia by manometric criteria). Using FLIP, there is a significant relationship with decreasing cross-sectional area associated with elevated BMI.[19]

All patients with typical symptoms of dysphagia, regurgitation, heartburn, or chest pain should raise suspicion for possible esophageal dysmotility and undergo endoscopic and manometric evaluations as described earlier. However, it is important to recognize that motility disorders may present inconspicuously in many patients, particularly in the obese patient population. Koppman et al. reported that in 116 obese patients (mean BMI 42.9), 41% had abnormal manometric findings, yet only one of these patients was symptomatic.[20] In a prospective study of 177 asymptomatic patients with morbid obesity who underwent HRM before metabolic surgery, HRM diagnosed esophageal motility disorder in 35.6% of patients.[21] This is particularly relevant in asymptomatic obese patients under consideration for bariatric surgery, as identification of a preexisting esophageal motility disorder has implications for potential postoperative complications, and thus surgical decision-making.[22] Therefore a systematic preoperative evaluation including HRM and/or adjunctive testing (barium esophagram or FLIP) should be considered for obese patients undergoing bariatric surgery, including asymptomatic patients.

Disorders of the esophagus

The most common disorder of the esophagus in obese patients includes EGJOO, IEM, DES, hypercontractile esophagus, and type II achalasia (Fig. 5.1). Prior definitions of EGJOO likely have led to an overdiagnosis of this condition that has been redefined by Chicago Classification 4.0.

For spastic disorders of the esophagus, there are currently three major subgroups: DES, hypercontractile esophagus (which includes the prior lexicon terminology of Jackhammer esophagus), and type III achalasia (will be discussed later this chapter). Spastic disorders usually have vast similarities with presentation; however, more recent changes in classification with the advent of HRM from conventional manometry along with updated diagnostic criteria from the Chicago classification have impacted how these disorders are defined.[23]

In general, the pathophysiology of spastic esophageal disorders has been incompletely understood. The current concept for the mechanism behind these disorders surrounds disturbances in the inhibitory and excitatory neuronal regulation of esophageal smooth muscles. The inhibitory and excitatory innervations involved in esophageal motor function of the longitudinal and circular muscle layers of the esophagus form the myenteric plexus. During the process of deglutition, inhibitory neurons are activated and lead to the release of nitric oxide (NO) and vasoactive intestinal peptide (VIP) resulting in relaxation of the esophageal body and the LES. Peristalsis is next with sequential contraction and relaxation of the smooth muscle as the bolus is propelled toward the lower esophageal sphincter. Overall, the excitatory innervation of the esophagus drives the basal tonicity of the esophageal body and LES, while the gradient from the inhibitory neurons initiate and promote peristalsis in the esophageal body and relaxation of the LES.[24]

Distal esophageal spasm and hypercontractile esophagus

Affecting approximately 1 in 100,000 individuals in the United States, DES is a rare motility disorder found in $<5\%$ of patients undergoing a motility evaluation for dysphagia.[25] DES is felt to be a defect in esophageal NO production associated with poor inhibitory innervation.[26] Prior studies have shown this alteration leads to dysfunction in distal contractile latency and in deglutitive inhibition with paired swallows.[27]

Hypercontractile esophagus has a set of very similar presenting symptoms to DES including chest pain/discomfort and dysphagia. It accounts for even less prevalence than DES, however, is thought to follow a similar physiologic pattern of excess cholinergic drive. Prior studies have supported this theory through measuring response to atropine.[28] Asynchrony between the longitudinal and circular muscle layers have been visualized on intraluminal ultrasound in individuals with nutcracker esophagus.[29]

Diagnosis

The assessment of esophageal disorders often starts with an upper endoscopy to evaluate for mechanical obstruction. Esophagogastroduodenoscopy (EGD) is often helpful in ruling out strictures, esophagitis, and esophageal rings that could be contributing to a patients' dysphagia. If these are absent on endoscopy, further testing to investigate for spastic disorders is often pursued.[24]

Esophageal manometry

HRM has become in the gold standard test for diagnosis of spastic esophageal disorders. Criteria for diagnosis are based on measurements of integrated relaxation pressure, distal latency, and distal contractile integral. Under the current Chicago Classification guidelines, DES diagnosis requires the presence of clinically relevant symptoms with a minimum of 20% premature contractions on HRM. Inconclusive diagnosis is identified by the presence of greater than 20% of simultaneous contractions in the distal esophagus during normal peristalsis with a distal latency less than 4.5 seconds and DCI of <450 mmHg.[23] To differentiate DES from achalasia, the Chicago classification requires a normal LES relaxation.[30]

Associated features seen with DES include spontaneous contractions, intermittent normal peristalsis, and multipeaked contractions. HRM will often show multipeaked contractions; however, this pattern can also occur with an esophagogastric junction outflow obstruction.[31]

Barium esophagram

This study is often used in addition to the previous interventions and can provide supporting data including esophagogastric junction (EGJ) bolus clearance, body peristalsis, and upper esophageal function. Peristaltic dysfunction is identified though incomplete esophageal emptying visualized as reversed movement of the bolus during the peristaltic wave. In a normal study, the EGJ will become patent when the bolus arrives. When there is suspicion for a spastic disorder on manometry, the classic finding of corkscrew appearance on esophagogram can be seen.[32] However, it has been shown that this modality, when used independently is not sufficient to diagnose nonachalasia motility disorders. Specifically, compared to the gold standard of HRM, barium esophagogram has shown limited sensitivity and specificity for detecting nonachalasia dysmotility.[33]

Treatment

In patients with hypercontractile motility disorders without GERD or well-controlled GERD, pharmacological therapy has been used, which includes a trial of calcium channel blockers (which reduce spastic contractions in the distal esophagus) or low dose tricyclic antidepressants (which diminish visceral sensory perception), though adverse effects may limit use and data in controlled trials have varied. Sublingual nitroglycerin or hyoscyamine as needed may also be beneficial in patients with mild and infrequent esophageal pain.[34]

Achalasia

Achalasia is a primary esophageal motility disorder that affects 10/100,000 individuals worldwide.[35] The peak incidence of achalasia is between 30 and 60 years old and occurs equally among men and women.[36–39] Similar to other disorders of esophageal motility, achalasia often presents with dysphasia to both solids and liquids along with regurgitation of undigested food. Obese patients have been seen to show symptoms of vomiting or choking on presentation.[40] Overall, the disease results from a lack of relaxation in the LES due to loss of inhibitory innervation.[41] The underlying cause of the loss of inhibitory neurons in achalasia still remains somewhat unclear; however, there are several proposed mechanisms.[42] There are currently three subtypes of achalasia (I, II, III) per the Chicago classification that are based on HRM. Weight loss can be variable dependent on the achalasia subtype with recent studies suggesting that type III achalasia was least likely and type II achalasia patients were most likely to have weight loss on presentation.[43]

The pathophysiology of achalasia is based on the selective loss or destruction of inhibitory neural fibers that subsequently results in the loss of esophageal peristalsis and lack of relaxation in the LES. In normal peristalsis, the balance between excitatory and inhibitory neurons driven by acetylcholine, NO and VIP lead to coordination of striated and smooth muscles within the esophagus.[44] The causes of initial insult to inhibitory neurons still remain unknown. It is suggested that preexisting autoimmune diseases or autoimmune triggers (viral infections) may play a role as inciting events.[41,45] Our group has previously shown the role of varicella zoster virus in the pathogenesis of achalasia.[46]

The hypothesized mechanism of progression to disease is thought to be mediated by progressive inflammation, leading to damage of the myenteric plexus and thus changes and loss of NO innervation. Chronic inflammation is thought to lead to autoimmunity, driving further inflammation and eventual loss of inhibitory innervation.[47,48]

Diagnosis

The diagnosis of achalasia can be accomplished by the following modalities.

Endoscopy

Similar to other esophageal motility disorders, patients with achalasia will often undergo endoscopy as initial diagnostic testing. This is important to rule out pseudoachalasia from possible malignancy or peptic stricture.[49] In obese patients with a prior laparoscopic gastric band, complications of this procedure can lead to a pseudoachalasia pattern, which is typically reversible with band deflation or band removal. Usual findings on EGD will show a puckered LES with retained food products and saliva. The esophageal body is usually dilated and might show complications of stasis such as candidiasis, ulcerations, or stasis esophagitis.[50]

Barium esophagram

This minimally invasive study will show a twisting or dilated esophagus with a tight LES with the classic "bird's beak" appearance. Esophageal emptying can be accessed via a timed barium esophagram.[44]

Esophageal manometry

This remains the primary modality of diagnosis for achalasia. A generalized diagnosis of achalasia is determined through demonstration of poor LES relaxation and impaired peristalsis without the presence of secondary esophageal obstruction (i.e., malignancy or tumor).[51] Subtypes of achalasia have been defined in the Chicago classification based on esophageal body pressurization patterns on HRM. All three phenotypes of achalasia will have lack of esophageal peristalsis and poor EGJ relaxation with elevated median IRP. In terms of smooth body contractility: for type I achalasia there is absent contractility; for type II achalasia there is absent contractility with 20% or more swallows with panesophageal pressurization, for type III achalasia there is presence of spasm without evidence of peristalsis.[23]

Treatment

In obese patients with concurrent achalasia, therapeutic options can be divided into the following categories: endoscopic (botulinum toxin injection, pneumatic dilation or PD, peroral endoscopic myotomy or POEM) and surgical. In the appropriate candidate, surgery via laparoscopic Heller myotomy (LHM) and Roux-en-Y gastric bypass is the preferred approach to treating achalasia in the morbidly obese patient, allowing treatment of both diseases simultaneously.[52] Roux-en-Y gastric bypass provides an additional antireflux advantage compared to alternative bariatric procedures such as laparoscopic sleeve gastrectomy, which can exacerbate preexisting GERD or give rise to de novo GERD.[53,54] In obese patients with achalasia who had undergone LHM with partial fundoplication (Dor procedure) without simultaneous bariatric surgery, many patients had symptom recurrence in long-term follow-up.[55]

POEM has been shown to have comparable clinical efficacy to LHM, particularly in patients with type III achalasia and distal esophageal spasm but has been associated with a higher incidence of postprocedure GERD compared to PD and LHM.[56,57] Obesity (BMI >30) is a positive predictive risk factor in the development of postoperative GERD after POEM.[58] Therefore obese patients with preexisting GERD or patients unwilling to use life-long proton pump inhibitor therapy should be strongly considered for surgical treatment, particularly LHM with Roux-en-Y gastric bypass. For patients of advanced age or at high surgical risk, a reasonable and less invasive option is graded pneumatic dilation, though esophageal perforation is a potential complication (incidence 1%–2%).[34] Lastly, Botulinum toxin injections, though effective in reducing LES pressure, have short-term benefit (approximately 6 months), and thus, should be reserved for high-risk patients with severe comorbidities precluding more invasive therapies or as a bridge to more definitive treatment (e.g., pregnant patients, patients on temporary dual-antiplatelet therapy.).[59] Notably, Botox injections also promote GERD by inducing hiatal (or crural diaphragm) paralysis.[60]

Ineffective esophageal motility

Ineffective esophageal motility is a nonspecific motility disorder identified by weak or failed peristalsis. It is the most common motility abnormality found on manometry (20%–30%); however, a standard diagnostic approach has been limited. There is limited knowledge on the pathophysiology of IEM; however, it is suspected that secondary peristalsis is impaired as this process is dependent on response to esophageal distention. The primary role of secondary peristalsis is to clear retained food, gastric reflux, air, liquid after failed primary peristalsis and is important to maintaining esophageal homeostasis.[61,62] Given that this process is impacted by vagal tone, it has been suggested that clinically relevant IEM and vagal dysfunction are intertwined.[63]

Multiple conditions have been shown to be associated with IEM including GERD and diabetes. One study assessing the prevalence of IEM in 89 patients with GERD found that 49.4% of patients met diagnostic criteria for IEM. Additionally, of patients with IEM, 67% had findings of esophagitis on EGD.[64] In regards to diabetes, poor glycemic control has been linked as a frequent etiology of esophageal dysmotility in patients with diabetic autonomic dysfunction.[65]

Diagnosis

On presentation, patients with IEM can be asymptomatic or have a wide array or symptoms similar to other motility disorders. An assessment of symptom distribution in 228 IEM patients showed that dysphagia was the most common presenting symptom in 25% of the patients. Additional common symptoms included cough (15%), chest pain (13%), heartburn (12%), and regurgitation (12%).[66] EGD is performed to rule out strictures, inflammation, and cancer (Fig. 5.4A).

Per the Chicago classification, a definitive diagnosis of IEM requires more than 70% ineffective swallows or at least 50% failed peristalsis (Fig. 5.4B).[23] The gold standard for diagnosis, as with many motility disorders, remains esophageal manometry. Intraluminal impedance monitoring has been shown to be an adjunct to manometry. Specifically, bolus clearance, transport and association with LES relaxation can be further delineated.[67] A study of 70 patients with IEM assessed esophageal bolus transit using combined impedance-manometry testing. Primary elements affecting bolus propagation in IEM patients were found to be amplitude of esophageal contraction <25 mmHg and the number of low amplitude

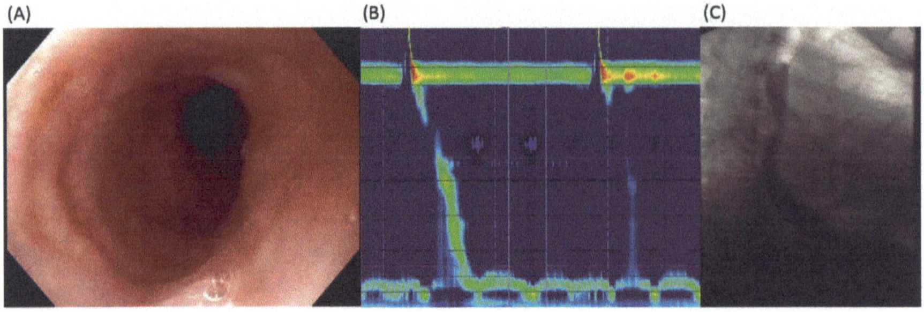

Figure 5.4 Ineffective motility with obesity (A) Upper endoscopy showing esophagus in a patient with morbid obesity, (B) High- resolution manometry (HRM) showing ineffective motility, (C) Barium esophagram showing poor bolus clearance.

swallows. 66% patients had abnormal bolus transport.[67] Esophagram can also be performed, which can show poor bolus clearance (Fig. 5.4C).

Treatment

Management of esophageal dysmotility is phenotype-directed and targeted to the individual patient, including patient preferences and comorbid diseases. HRM cannot determine which patients have symptomatic reflux and hence ambulatory esophageal pH monitoring can be helpful to determine if these patients have pathological GERD. In obese patients with both esophageal motility disorder and associated GERD, adequate treatment with proton pump inhibitor therapy is the first step. Dietary and lifestyle modifications can also promote weight loss and reduce symptom burden. Unfortunately, there are no current medical therapy that can restore esophageal peristalsis to reverse the manometric findings of IEM. The 5-HT4 agonist have shown to improve bolus transport by increasing esophageal contractility, but these medications have limited by availability, side-effect profile, and reproducibility in human trials.[68,69] For patients with GERD-related complications, weight loss surgery can be offered as IEM is not a contraindication for these surgeries.[70]

Conclusion

Esophageal dysmotility and GERD are common in patients with obesity, but many patients do not have symptoms. Recognition of these motility disorders is important in the long-term management of this patients. HRM and EGD are commonly used tools to help delineate functional motility pathology. Newer advances in FLIP may provide an opportunity to better understand distensibility in relationship to obesity. Given the rates of dysmotility postbariatric surgery, early recognition before bariatric approaches is also important to provide personalized care to this important patient population.

References

1. Remus Popa A, Fratila O, Rus M, et al. Risk factors for adiposity in the urban population and influence on the prevalence of overweight and obesity. *Exp Ther Med.* 2020;20(1):129–133.
2. Nguyen NT, Magno CP, Lane KT, Hinojosa MW, Lane JS. Association of hypertension, diabetes, dyslipidemia, and metabolic syndrome with obesity: findings from the National Health and Nutrition Examination Survey, 1999 to 2004. *J Am Coll Surg.* 2008;207(6):928–934.
3. Popescu AL, Costache RS, Costache DO, et al. Manometric changes of the esophagus in morbidly obese patients. *Exp Ther Med.* 2021;21(6):604.
4. Cote-Daigneault J, Leclerc P, Joubert J, Bouin M. High prevalence of esophageal dysmotility in asymptomatic obese patients. *Can J Gastroenterol Hepatol.* 2014;28(6):311–314.
5. Jaffin BW, Knoepflmacher P, Greenstein R. High prevalence of asymptomatic esophageal motility disorders among morbidly obese patients. *Obes Surg.* 1999;9(4):390–395.
6. El-Serag HB, Graham DY, Satia JA, Rabeneck L. Obesity is an independent risk factor for GERD symptoms and erosive esophagitis. *Am J Gastroenterol.* 2005;100(6):1243–1250.
7. Wu JC, Mui LM, Cheung CM, Chan Y, Sung JJ. Obesity is associated with increased transient lower esophageal sphincter relaxation. *Gastroenterology.* 2007;132(3):883–889.
8. Naik RD, Choksi YA, Vaezi MF. Consequences of bariatric surgery on oesophageal function in health and disease. *Nat Rev Gastroenterol Hepatol.* 2016;13(2):111–119.
9. Cote-Daigneault J, Poitras P, Rabasa-Lhoret R, Bouin M. Plasma leptin concentrations and esophageal hypomotility in obese patients. *Can J Gastroenterol Hepatol.* 2015;29(1):49–51.
10. Fox MR, Sweis R, Yadlapati R, et al. Chicago classification version 4.0 © technical review: update on standard high-resolution manometry protocol for the assessment of esophageal motility. *Neurogastroenterol Motil.* 2021;33(4).
11. Rohof WOA, Bredenoord AJ. Chicago classification of esophageal motility disorders: lessons learned. *Curr Gastroenterol Rep.* 2017;19(8).
12. Yadlapati R, Kahrilas PJ, Fox MR, et al. Esophageal motility disorders on high-resolution manometry: Chicago classification version 4.0 ©. *Neurogastroenterol Motil.* 2021;33(1).
13. Yen H-H, Tseng P-H, Shih M-C, Yang P-J, Lin M-T, Lee P-C. Derangement of esophageal anatomy and motility in morbidly obese patients: a prospective study based on high-resolution impedance manometry. *Surg Obes Relat Dis.* 2020;16(12):2006–2015.
14. Mora F, Cassinello N, Mora M, Bosca M, Minguez M, Ortega J. Esophageal abnormalities in morbidly obese adult patients. *Surg Obes Relat Dis.* 2016;12(3):622–628.
15. Santonicola A, Angrisani L, Vitiello A, et al. Hiatal hernia diagnosis prospectively assessed in obese patients before bariatric surgery: accuracy of high-resolution manometry taking intraoperative diagnosis as reference standard. *Surg Endosc.* 2020;34(3):1150–1156.
16. Carlson DA, Pandolfino JE. High-resolution manometry in clinical practice. *Gastroenterol Hepatol (NY).* 2015;11(6):374–384.
17. Desprez C, Roman S, Leroi AM, Gourcerol G. The use of impedance planimetry (Endoscopic Functional Lumen Imaging Probe, EndoFLIP(®)) in the gastrointestinal tract: a systematic review. *Neurogastroenterol Motil.* 2020;32(9):e13980.

18. Donnan EN, Pandolfino JE. EndoFLIP in the esophagus: assessing sphincter function, wall stiffness, and motility to guide treatment. *Gastroenterol Clin North Am*. 2020;49(3):427–435.
19. Tucker E, Sweis R, Anggiansah A, et al. Measurement of esophago-gastric junction cross-sectional area and distensibility by an endolumenal functional lumen imaging probe for the diagnosis of gastro-esophageal reflux disease. *Neurogastroenterol Motil*. 2013;25(11):904–910.
20. Koppman JS, Poggi L, Szomstein S, Ukleja A, Botoman A, Rosenthal R. Esophageal motility disorders in the morbidly obese population. *Surg Endosc*. 2007;21(5):761–764.
21. Kristo I, Paireder M, Jomrich G, et al. Silent gastroesophageal reflux disease in patients with morbid obesity prior to primary metabolic surgery. *Obes Surg*. 2020;30 (12):4885–4891.
22. Braghetto I, Lanzarini E, Korn O, Valladares H, Molina JC, Henriquez A. Manometric changes of the lower esophageal sphincter after sleeve gastrectomy in obese patients. *Obes Surg*. 2010;20(3):357–362.
23. Yadlapati R, Kahrilas PJ, Fox MR, et al. Esophageal motility disorders on high-resolution manometry: Chicago classification version 4.0 ©. *Neurogastroenterol Motil*. 2021;33(1):e14058.
24. Roman S, Kahrilas PJ. Management of spastic disorders of the esophagus. *Gastroenterol Clin North Am*. 2013;42(1):27–43.
25. Sperandio M, Tutuian R, Gideon RM, Katz PO, Castell DO. Diffuse esophageal spasm: not diffuse but distal esophageal spasm (DES). *Dig Dis Sci*. 2003;48(7):1380–1384.
26. Orlando RC, Bozymski EM. Clinical and manometric effects of nitroglycerin in diffuse esophageal spasm. *N Engl J Med*. 1973;289(1):23–25.
27. Behar J, Biancani P. Pathogenesis of simultaneous esophageal contractions in patients with motility disorders. *Gastroenterology*. 1993;105(1):111–118.
28. Hong YS, Min YW, Rhee PL. Two distinct types of hypercontractile esophagus: classic and spastic jackhammer. *Gut Liver*. 2016;10(5):859–863.
29. Jung HY, Puckett JL, Bhalla V, et al. Asynchrony between the circular and the longitudinal muscle contraction in patients with nutcracker esophagus. *Gastroenterology*. 2005;128(5):1179–1186.
30. Bredenoord AJ, Fox M, Kahrilas PJ, et al. Chicago classification criteria of esophageal motility disorders defined in high resolution esophageal pressure topography. *Neurogastroenterol Motil*. 2012;24(Suppl 1):57–65.
31. Spechler SJ, Castell DO. Classification of oesophageal motility abnormalities. *Gut*. 2001;49(1):145–151.
32. Pandolfino JEKP. Esophageal neuromuscular function and motility disorders. In: Feldman MFL, Brandt L, eds. *Sleisenger and Fordtran's Gastrointestinal and Liver Disease*. 10th ed Saunders; 2016:733–754. e8.
33. Finnerty BM, Aronova A, Cheguevara A, et al. Esophageal dysmotility and the utility of barium swallow: an opaque diagnosis. *Gastroenterology*. 2015;148:S1131–S1132.
34. Mittal R, Vaezi MF. Esophageal motility disorders and gastroesophageal reflux disease. *N Engl J Med*. 2020;383(20):1961–1972.
35. Sadowski DC, Ackah F, Jiang B, Svenson LW. Achalasia: incidence, prevalence and survival. A population-based study. *Neurogastroenterol Motil*. 2010;22(9):e256–e261.
36. Francis DL, Katzka DA. Achalasia: update on the disease and its treatment. *Gastroenterology*. 2010;139(2):369–374.
37. Sonnenberg A. Hospitalization for achalasia in the United States 1997–2006. *Dig Dis Sci*. 2009;54(8):1680–1685.
38. Sonnenberg A, Massey BT, McCarty DJ, Jacobsen SJ. Epidemiology of hospitalization for achalasia in the United States. *Dig Dis Sci*. 1993;38(2):233–244.
39. Vaezi MF, Richter JE. Diagnosis and management of achalasia. American college of gastroenterology practice parameter committee. *Am J Gastroenterol*. 1999;94(12):3406–3412.

40. Rakita SS, Villadolid D, Kalipersad C, Thometz D, Rosemurgy A. BMI affects presenting symptoms of achalasia and outcome after Heller myotomy. *Surg Endosc.* 2007;21(2):258–264.
41. Boeckxstaens GE. Achalasia: virus-induced euthanasia of neurons? *Am J Gastroenterol.* 2008;103(7):1610–1612.
42. Frieling T, Berges W, Borchard F, Lubke HJ, Enck P, Wienbeck M. Family occurrence of achalasia and diffuse spasm of the oesophagus. *Gut.* 1988;29(11):1595–1602.
43. Patel DA, Naik R, Slaughter JC, Higginbotham T, Silver H, Vaezi MF. Weight loss in achalasia is determined by its phenotype. *Dis Esophagus.* 2018;31(9).
44. Patel DA, Lappas BM, Vaezi MF. An overview of achalasia and its subtypes. *Gastroenterol Hepatol (NY).* 2017;13(7):411–421.
45. Booy JD, Takata J, Tomlinson G, Urbach DR. The prevalence of autoimmune disease in patients with esophageal achalasia. *Dis Esophagus.* 2012;25(3):209–213.
46. Naik RD, Vaezi MF, Gershon AA, et al. Association of achalasia with active varicella zoster virus infection of the esophagus. *Gastroenterology.* 2021;161(2):719–721. e2.
47. Ates F, Vaezi MF. The pathogenesis and management of achalasia: current status and future directions. *Gut Liver.* 2015;9(4):449–463.
48. Kahrilas PJ, Boeckxstaens G. The spectrum of achalasia: lessons from studies of pathophysiology and high-resolution manometry. *Gastroenterology.* 2013;145(5):954–965.
49. Kahrilas PJ, Kishk SM, Helm JF, Dodds WJ, Harig JM, Hogan WJ. Comparison of pseudoachalasia and achalasia. *Am J Med.* 1987;82(3):439–446.
50. Patel DA, Kim HP, Zifodya JS, Vaezi MF. Idiopathic (primary) achalasia: a review. *Orphanet J Rare Dis.* 2015;10:89.
51. Pandolfino JE, Kwiatek MA, Nealis T, Bulsiewicz W, Post J, Kahrilas PJ. Achalasia: a new clinically relevant classification by high-resolution manometry. *Gastroenterology.* 2008;135(5):1526–1533.
52. Ithurralde-Argerich J, Rosner L, Faerberg A, Puma R, Ferro D, Cuenca-Abente F. Laparoscopic Heller Myotomy and Roux-en-Y gastric bypass as treatment for patients with achalasia and morbid obesity: outcomes in a short series of patients. *J Laparoendosc Adv Surg Tech A.* 2021;31(1):29–35.
53. Naik RD, Choksi YA, Vaezi MF. Impact of weight loss surgery on esophageal physiology. *Gastroenterol Hepatol (NY).* 2015;11(12):801–809.
54. Fisichella PM, Orthopoulos G, Holmstrom A, Patti MG. The surgical management of achalasia in the morbid obese patient. *J Gastrointest Surg.* 2015;19(6):1139–1143.
55. Harrison JM, Rakestraw SL, Doane SM, Pucci MJ, Palazzo F, Chojnacki KA. Achalasia and obesity: patient outcomes and impressions following laparoscopic Heller myotomy and Dor fundoplication. *Langenbecks Arch Surg.* 2020;405(6):809–816.
56. Khashab MA, Vela MF, Thosani N, et al. ASGE guideline on the management of achalasia. *Gastrointest Endosc.* 2020;91(2):213–227. e6.
57. Repici A, Fuccio L, Maselli R, et al. GERD after per-oral endoscopic myotomy as compared with Heller's myotomy with fundoplication: a systematic review with *meta*-analysis. *Gastrointest Endosc.* 2018;87(4):934–943. e18.
58. Ward MA, Whitfield EP, Hasan SS, Ogola GO, Leeds SG. Objectively confirmed gastroesophageal reflux disease following per oral endoscopic myotomy higher in obese patients (BMI > 30). *Surg Laparosc Endosc Percutan Tech.* 2020;31(2):146–149.
59. Oude Nijhuis RAB, Prins LI, Mostafavi N, van Etten-Jamaludin FS, Smout AJPM, Bredenoord AJ. Factors associated with achalasia treatment outcomes: systematic review and *meta*-analysis. *Clin Gastroenterol Hepatol.* 2020;18(7):1442–1453.
60. Kumar D, Zifan A, Mittal RK. Botox injection into the lower esophageal sphincter induces hiatal paralysis and gastroesophageal reflux. *Am J Physiol Gastrointest Liver Physiol.* 2020;318(1):G77–g83.

61. Chen CL, Yi CH, Liu TT. Relevance of ineffective esophageal motility to secondary peristalsis in patients with gastroesophageal reflux disease. *J Gastroenterol Hepatol.* 2014;29(2):296–300.
62. Helm JF, Dodds WJ, Pelc LR, Palmer DW, Hogan WJ, Teeter BC. Effect of esophageal emptying and saliva on clearance of acid from the esophagus. *N Engl J Med.* 1984;310(5):284–288.
63. Chen JH. Ineffective esophageal motility and the vagus: current challenges and future prospects. *Clin Exp Gastroenterol.* 2016;9:291–299.
64. Ho SC, Chang CS, Wu CY, Chen GH. Ineffective esophageal motility is a primary motility disorder in gastroesophageal reflux disease. *Dig Dis Sci.* 2002;47(3):652–656.
65. Ahmed W, Vohra EA. Esophageal motility disorders in diabetics with and without neuropathy. *J Pak Med Assoc.* 2006;56(2):54–58.
66. Abdel Jalil AA, Castell DO. Ineffective esophageal motility (IEM): the old-new frontier in esophagology. *Curr Gastroenterol Rep.* 2016;18(1):1.
67. Tutuian R, Castell DO. Clarification of the esophageal function defect in patients with manometric ineffective esophageal motility: studies using combined impedance-manometry. *Clin Gastroenterol Hepatol.* 2004;2(3):230–236.
68. Staiano A, Clouse RE. The effects of cisapride on the topography of oesophageal peristalsis. *Aliment Pharmacol Ther.* 1996;10(6):875–882.
69. Shetler KP, Bikhtii S, Triadafilopoulos G. Ineffective esophageal motility: clinical, manometric, and outcome characteristics in patients with and without abnormal esophageal acid exposure. *Dis Esophagus.* 2017;30(6):1–8.
70. Schneider R, Lazaridis I, Kraljevic M, Beglinger C, Wolnerhanssen B, Peterli R. The impact of preoperative investigations on the management of bariatric patients; results of a cohort of more than 1200 cases. *Surg Obes Relat Dis.* 2018;14(5):693–699.

CHAPTER 6

Nonsurgical management of GERD in obesity

Okeefe L. Simmons[1,2], Rekha B. Kumar[2] and Gitanjali Srivastava[3,4,5]
[1]Division of Gastroenterology, Weill Cornell Medical College, New York, NY, United States
[2]Division of Endocrinology, Diabetes & Metabolism, Weill Cornell Medical College, New York, NY, United States
[3]Division of Diabetes, Endocrinology & Metabolism, Department of Medicine, Vanderbilt Weight Loss Center, Vanderbilt University School of Medicine, Nashville, TN, United States
[4]Department of Surgery, Vanderbilt University School of Medicine, Nashville, TN, United States
[5]Department of Pediatrics, Vanderbilt University School of Medicine, Nashville, TN, United States

Introduction

The diagnosis and management of gastroesophageal reflux disease (GERD) in patients with obesity is similar to patients without obesity, and weight loss has been shown to improve GERD.[1] Medical obesity therapies, such as antiobesity pharmacotherapy, and bariatric surgery are unique treatment considerations in patients with obesity presenting with GERD symptoms. The initial management of GERD is centered around conservative therapeutic options. Counseling patients about diet, lifestyle changes, and pharmacotherapy are important aspects in managing GERD. The goal of treatment is symptomatic relief and the prevention of complications such as esophageal adenocarcinoma, which occurs more commonly in patients with obesity.[2,3]

Diet and lifestyle therapy

Diet

A dietary history focused on the type of food/beverages consumed, and meal times in relation to sleep may be helpful in identifying the role of diet in the exacerbation of GERD symptoms. Although studies have yielded mixed results, clinical guidelines suggest the avoidance of food and beverages associated with exacerbation of GERD symptoms.[4,5] These specific triggers may differ among patients, and thus the recommendations should be individualized for each patient.[6] Commonly reported culprits have been proposed to induce symptoms through various mechanisms

Obesity and Esophageal Disorders
DOI: https://doi.org/10.1016/B978-0-323-98365-5.00002-4

including decreased lower esophageal sphincter (LES) tone (chocolate, coffee, alcohol), increased transient lower esophageal sphincter relaxations (TLESRs) (carbonated beverages), and direct mucosal irritation (citrus, spicy foods).[4,7,8]

No formal recommendations regarding optimal dietary macronutrient composition (i.e., low fat, low carbohydrate, or high protein) are available, as studies have yielded mixed results.[2,8–10] The most robust evidence exists in support of a low carbohydrate diet to reduce GERD symptoms.[11–13] A prospective study of 144 patients with GERD reported symptomatic improvement after reducing carbohydrate intake for 2 weeks.[12]

The relationship between timing of meals and GERD symptoms has also been explored. Supine positioning and sleep after meals are hypothesized to increase postprandial reflux. Although results of pH studies and patient symptom surveys that compared early versus late-night meals have produced conflicting results, most prospective data found an association between late-night eating and worsening GERD symptoms.[7,14,15] Therefore patients should be advised to avoid recumbency 3 hours after eating.[5,15]

Lifestyle

Lifestyle changes may alleviate or eliminate GERD symptoms. Weight loss should be encouraged and regularly assessed among patients with GERD and obesity. Weight loss in patients with obesity and overweight and in individuals with normal weight who experienced recent weight gain decreases the prevalence of GERD and improves symptoms.[1,16–18] At least 5% weight loss for women and at least 10% weight loss for men is recommended for GERD resolution.[1] A prospective study of 332 patients with overweight or obesity enrolled in a structured weight loss program showed a dose–response correlation between percent weight loss and resolution of GERD symptoms at 6 months. In this study, over 80% of patients experienced symptomatic improvement and nearly two-thirds had complete resolution of GERD following weight reduction.[1] Obesity is associated with GERD and its complications—erosive esophagitis (EE), Barretts esophagus (BE), and esophageal adenocarcinoma.[3,19–22] Thus weight loss and adequate control of GERD are of particular importance in these patients. A referral to an obesity medicine specialist should be considered to assist with weight loss in patients with obesity, especially for patients having difficulty achieving weight loss.

Tobacco smoking cessation and avoiding excess alcohol intake as general health measures should be encouraged. Studies examining the association of these behaviors with GERD have not produced consistent results.[22–24] However, larger studies and a number of recent studies have demonstrated an association between smoking tobacco and GERD. A study of over 7000 patients with GERD identified tobacco use (current or former use of at least 11 cigarettes daily) and regular consumption of spirits as risk factors for worsening GERD symptoms.[22] Furthermore, multicenter studies and *meta*-analyses have identified smoking as a risk factor for both BE and esophageal adenocarcinoma.[25–27]

Head of the bed elevation is associated with an improvement in GERD symptoms.[28,29] It is hypothesized that in the supine position during sleep there is prolonged reflux of acidic gastric contents when compared to the upright position.[30] A study that assessed patient-reported symptoms as well as objective measures of nocturnal acid exposure among patients with nocturnal GERD revealed significant symptomatic improvement and decreased esophageal acid exposure with 20 cm elevation of the head of the bed.[29] Guidelines for GERD management recommend the head of the bed elevation, which can be accomplished using blocks or a wedge; this recommendation may be of particular importance for patients with nocturnal GERD symptoms.[5]

Acid-suppression pharmacotherapy

Proton pump inhibitor

The cornerstone of pharmacologic management of GERD is proton pump inhibitors (PPIs) (Table 6.1). PPIs irreversibly inhibit the $H+/K+$ ATPase ("proton pump") of parietal cells in the proximal stomach, which decreases gastric acid production. However, gastric acid production is often normal, so PPIs do not usually address the underlying pathophysiologic process in GERD. By increasing the duration of time that the acidic refluxate pH is 4 or greater, PPI use results in symptomatic improvement of GERD.[35]

PPIs are first-line therapy in patients with typical GERD. In patients with typical uncomplicated GERD (i.e., no evidence of EE or BE), a trial of daily PPI is initiated and continued for 4 weeks.[5,36] If there is symptomatic resolution, an attempt should be made to discontinue PPI after 4–8 weeks. If maintenance therapy is required, switching to an alternative class of medications such as a histamine$_2$ receptor antagonist (H_2RA) should be encouraged. In a prospective study of patients with PPI-responsive

Table 6.1 Pharmacotherapy for GERD management.

Medication class	Mechanism of action	Indication(s)
Proton pump inhibitor (e.g., omeprazole, pantoprazole, esomeprazole, lansoprazole, rabeprazole, dexlansoprazole)	Suppress acid production by irreversible inhibition of the H + / K + ATPase ("proton pump") of parietal cells in the proximal stomach	• First line therapy • Maintenance therapy for Barret's esophagus, erosive esophagitis
Histamine$_2$ receptor antagonist (e.g., famotidine, nizatidine, cimetidine)	Suppress acid production by reversible inhibition of parietal cell histamine$_2$ receptors in the proximal stomach	• Step-down maintenance therapy • Second line option when PPI use is contraindicated • Poorly controlled nocturnal GERD
Potassium competitive acid blocker[a] (e.g., vonoprazan)	Suppress acid production by reversible competitive inhibition of the potassium binding site of the parietal cell H + /K + ATP-ase in the proximal stomach[31]	• Alternative option for typical GERD, erosive esophagitis
GABA-B receptor agonist (e.g., baclofen)	Decreases TLESR via stimulation of GABA-B receptors both: (1) Centrally—stimulation of neurons of the motor nucleus of the vagal nerve and nucleus tract solitarious and (2) Peripherally—inhibition of TSELRs due to esophageal distention[32]	• PPI refractory GERD • Use with caution due to possible side effects
Antacid (e.g., aluminum hydroxide, calcium carbonate)	Relieve symptoms by neutralizing acid present in the stomach	• Immediate symptomatic relief if infrequent symptoms
Alginate (e.g., Gaviscon)	Precipitates upon contact with the postprandial acid pocket in the	• PPI refractory uncomplicated GERD

(*Continued*)

Table 6.1 (Continued)

Medication class	Mechanism of action	Indication(s)
	proximal stomach to form a mechanical barrier ("raft") between the acidic gastric refluxate and esophageal mucosa[33]	
Sucralfate	Local barrier protection against acid-induced mucosal damage	• Consider use in pregnancy
Prokinetic 1. (e.g., metoclopramide, erythromycin, domperidone[a])	Various mechanisms (e.g., 5-HT_3 antagonist, 5-HT_4 agonist, $dopamine_2$ antagonist, $alpha_2$-adrenergic antagonist), which increase LES tone, increase strength of esophageal contractions, and accelerate gastric emptying[34]	• Patients with documented gastroparesis • Caution with central acting agents (e.g., metoclopramide)

GABA, Gamma-aminobutyric acid; *LES*, lower esophageal sphincter; *PPI*, proton pump inhibitor; *TLESR*, transient lower esophageal sphincter relaxation.
[a]Not available in the United States.

GERD, at 1 year 58% of patients gradually discontinued PPI, approximately one-third of the study cohort were successfully maintained on an H_2RA only, and 15% discontinued all GERD pharmacotherapy.[37]

In patients unable to discontinue PPI, compliance should be assessed as over 50% of patients take PPIs incorrectly.[38] Ensuring that patients are taking their medication as directed is an important issue that should be constantly reassessed, as to avoid unnecessary dose escalation or unwarranted additional evaluation. Furthermore, when assessing compliance it is important to confirm that patients prescribed a delayed-release PPI are taking it 30−60 minutes before a meal.[5,39]

For patients with incomplete relief of GERD symptoms despite daily PPI adherence, different management strategies have been studied. Increasing the frequency from daily to twice daily for an additional 8 weeks is recommended in GERD management guidelines.[5,40] Split dosing (e.g., twice daily rather than daily) increases the duration of time that refluxate pH is 4 or greater.[41]

After any dose escalation, attempts should be made to decrease the dose and/or frequency of administration according to patient response to avoid PPI overuse.[42] Another proposed strategy is to switch to a different type of PPI due to differences in PPI metabolism. PPIs can be classified as CYPC219 dependent or independent based on differences in metabolism due to a genetic CYP isoenzyme polymorphism (CYPC219).[43,44] Drug levels of CYPC219-dependent PPIs can vary between patients as a result.[45] Use of a patient's genotype to tailor therapy (PPI type and/or dosing), or switching to a CYPC219 independent PPI (e.g., Rabeprazole) have been proposed as methods to increase PPI response.[44] Although a *meta*-analysis demonstrated no efficacy among the different PPIs, emerging data suggest that switching PPIs may be an effective option for patients with partial response to PPIs.[35]

Patients with PPI refractory GERD—persistent symptoms despite PPI optimization and a confirmed diagnosis of GERD—should undergo additional evaluation.[5,46] Alternative diagnoses such as functional heartburn, reflux hypersensitivity, eosinophilic esophagitis, and rumination syndrome should be considered and managed appropriately if identified. Refractory symptoms exist due to failure of the protective anatomic or physiologic antireflux mechanisms: decreased antireflux barrier (including LES and crural diaphragm) function, diminished esophageal acid clearance, impaired esophageal mucosal integrity (leading to increased mucosal permeability), delayed gastric emptying, or increased TLESRs.[46,47] Furthermore, identification of the specific antireflux component or alternative diagnosis can guide the decision regarding which adjunct agent to trial as monotherapy or in combination with acid suppression.[46,47]

EE and BE are GERD complications managed with maintenance PPI administration. Twice-daily PPI administration for 8 weeks is recommended initially for EE healing, followed by a maintenance daily PPI to promote continued remission.[5,48] A prospective study comparing the efficacy of treatment strategies demonstrated 80% remission of EE with maintenance PPI at 1 year, whereas less than half of the patients on maintenance H2RA were in remission during follow-up endoscopy.[48] Daily maintenance PPI is also recommended for patients with BE to decrease the risk of dysplasia and progression to esophageal adenocarcinoma.[49] A *meta*-analysis demonstrated that maintenance PPI use in patients with BE reduced the risk of progression to high-grade dysplasia and/or esophageal adenocarcinoma by 71%.[50] Maintenance PPI should be taken once daily; there is no additional proven preventative benefit of twice-daily PPI for asymptomatic patients with BE.[51–53]

PPIs are commonly prescribed for extraesophageal symptoms attributed to GERD (e.g., noncardiac chest pain, chronic cough, asthma, laryngitis). However, multiple studies demonstrated a lack of efficacy when PPIs were used for these symptoms. Fewer than 15% of patients with noncardiac chest pain without objective evidence of GERD respond to PPI.[54] Separate Cochrane systematic reviews did not find conclusive evidence to support PPI efficacy in the treatment of asthma or chronic cough.[55,56] Patients with nonurgent extraesophageal manifestations and typical GERD symptoms should consider a trial of PPI therapy; those without improvement should be referred for further evaluation to assess for alternate causes of extraesophageal symptoms.[5,57]

PPIs use has been associated with various conditions, including *Clostridium difficile* infection, osteoporosis, kidney disease, cancer, and community-acquired pneumonia; however, bias is an important limitation of these studies that were predominantly observational.[36] A large randomized controlled trial of 17,589 patients demonstrated a possible increased risk of enteric infections (OR 1.33; 95% CI: 1.01–1.75) associated with PPI use, but otherwise no difference in safety between Pantoprazole 40 mg daily and placebo.[58] The American College of Gastroenterology recommends caution with PPI use in patients at risk for *Clostridium difficile* infection.[5] In pregnancy, one *meta*-analysis suggested an association between maternal PPI use and congenital malformation; however, the authors noted multiple notable limitations that impacted study quality, including clinical heterogeneity, as well as selection bias and recall bias.[59,60] Furthermore, other *meta*-analyses have failed to demonstrate a significant association between PPI use and congenital malformations.[61,62] A discussion of the risks and benefits of PPI use during pregnancy should occur when clinically indicated.[5]

Histamine$_2$ receptor antagonist

H_2RAs suppress acid production by competitive reversible inhibition of parietal cell histamine$_2$ receptors in the proximal stomach; however, acetylcholine and gastrin-mediated parietal cell acid production continues.[63] As a result, H_2RAs are less effective than PPIs for acid suppression.[64] The relative risk of heartburn remission in a Cochrane systematic review was 0.37 (95% CI: 0.32–0.44) versus 0.77 (95% CI: 0.60–0.99) for H_2RAs in placebo-controlled trials.[65] H_2RAs offer an alternative to patients who are unable to tolerate PPIs or when PPI use is contraindicated.

H_2RAs are used in step-down therapy when attempting to wean patients the PPI-responsive GERD symptoms off of PPI. Patients who respond to PPI therapy can trial a course of an H_2RA up to twice daily to control symptoms. No benefit to H_2RA therapy administered more than twice daily has been demonstrated. Nighttime H_2RA can be considered for symptomatic patients with objective evidence of nocturnal GERD not adequately controlled with PPIs, as combination therapy of H_2RA with PPIs enhances nocturnal gastric acid suppression.[5,66,67]

Importantly, H_2RA monotherapy is not recommended for patients with EE or BE.[5,51] A *meta*-analysis demonstrated that after 2–12 weeks of therapy, significantly more healing of EE was seen in patients on PPI (83.6%) then H_2RA (51.9%).[68] Unlike the data seen in maintenance PPI therapy, maintenance H_2RA does not decrease the risk of high-grade dysplasia or esophageal adenocarcinoma.[49,51]

H_2RA therapy is well tolerated by most patients. Tachyphylaxis may occur within weeks of H_2RA use.[5,69] Due to increased levels of the human carcinogen *N*-nitrosodimethylamine, Ranitidine was withdrawn from the market by the United States Food and Drug Administration.[70]

Potassium competitive acid blocker

Although they are not available in the United States, the potassium competitive acid blocker (PCAB) vonoprazan is approved for use in Japan and South Korea. Acid suppression with PCABs occurs via reversible competitive inhibition of the potassium binding site of the $H+/K+$ ATP-ase. A *meta*-analysis of 42 studies revealed similar GERD healing with vonoprazan compared to most PPIs, and superior EE healing with vonoprazan compared to all studied PPIs at 8 weeks.[31]

Table 6.1 provides a summary of pharmacological management of GERD.

Other pharmacotherapy

Patients with GERD who do not have adequate symptom control with lifestyle change and acid suppression therapy may require an adjunct medication. These medications—gamma-aminobutyric acid (GABA)-B receptor agonists, antacids, alginates, sucralfate, and prokinetics—have not been shown to be superior in relieving GERD symptoms compared to PPIs.[5,65,71,72] Furthermore, adjunct medications have no role in the management of asymptomatic EE or BE.[5,48]

TLESRs are a cause of PPI refractory GERD. Baclofen is a GABA-B receptor agonist, which decreases TLESRs. A randomized placebo-controlled trial demonstrated decreased esophageal acid exposure time and fewer reflux episodes in patients who received Baclofen.[32] Baclofen use is most commonly limited by central nervous system side effects, including somnolence and dizziness.

Antacids contain salts of magnesium, aluminum, calcium, or sodium. Although they provide immediate relief of symptoms by acid neutralization, the duration of symptomatic relief is short so they should not be used by patients with frequent symptoms.[73,74] Alginates are derivates of alginic acids that precipitate upon contact with the postprandial acid pocket. This precipitate localizes to the acid pocket and provides a mechanical barrier between the acidic gastric refluxate and esophageal mucosa.[33] Alginate-containing compounds, often in combination with antacids (e.g., Gaviscon), improve nonerosive GERD symptoms when compared to placebo or antacids alone.[33,75]

Sucralfate is a complex of sucrose sulfate and aluminum hydroxide that provides local barrier protection against acid-induced mucosal damage.[76] Symptomatic relief with sucralfate is superior to placebo in nonerosive GERD; however, it must be taken up to four times daily and has not shown benefit compared to acid-suppressing agents.[68,72,76] Thus the American College of Gastroenterology does not recommend routine use of sucralfate for GERD management in patients who are not pregnant.[5]

Prokinetics (e.g., metoclopramide, cisapride, erythromycin, domperidone) increase LES tone and the strength of esophageal contractions.[34] However, no consistent clinical benefit has been shown in patients with GERD.[68] Prokinetics may have a role in patients with GERD who also have objective evidence of gastroparesis.[5]

Neuromodulators (tricyclic antidepressants, selective serotonin reuptake inhibitors, serotonin-norepinephrine reuptake inhibitors, trazodone) are important therapeutic options in patients with a functional esophageal disorder, which can be misdiagnosed as GERD or contribute to symptoms along with GERD. If a functional esophageal disorder (e.g., functional heartburn, reflux hypersensitivity) with or without accompanying GERD is identified during the workup of PPI refractory GERD, a neuromodulator can be used as monotherapy or in combination with a PPI if indicated.[71]

Summary

Proton pump inhibitors are the mainstay of pharmacotherapy for typical GERD and complicated GERD (e.g., EE and BE). Symptoms should be regularly reassessed to determine if step-down therapy or discontinuation of acid suppression therapy is possible for patients with uncomplicated GERD. Patients with PPI refractory GERD should undergo additional workup and may ultimately require the use of an adjunct medication. Importantly, dietary and lifestyle changes, with an emphasis on the importance of weight loss, should accompany any pharmacologic therapy recommended for the nonsurgical management of GERD in patients with obesity. Patients should be referred to an obesity medicine specialist for individualized lifestyle recommendations and initiation of antiobesity medication to achieve target weight loss (≥5% for women, ≥10% for men) for symptomatic relief. Referral to a surgeon for antireflux and/or bariatric surgical options is indicated for a subset of patients with refractory symptoms or additional obesity-related comorbidities.

Disclosures

OLS has nothing to disclose. RBK reports receiving lecture fees from Novo Nordisk and Janssen Pharmaceuticals, consulting fees from Gelesis, Pfizer, and Eli Lilly, and having an equity interest in Vivus. GS reports receiving lecture fees from Novo Nordisk and Rhythm Pharmaceuticals and consulting fees from Eli Lilly.

References

1. Singh M, Lee J, Gupta N, et al. Weight loss can lead to resolution of gastroesophageal reflux disease symptoms: a prospective intervention trial. *Obes (Silver Spring)*. 2013;21(2):284–290.
2. El-Serag HB, Satia JA, Rabeneck L. Dietary intake and the risk of gastro-oesophageal reflux disease: a cross sectional study in volunteers. *Gut*. 2005;54(1):11–17.
3. Hampel H, Abraham NS, El-Serag HB. *Meta*-analysis: obesity and the risk for gastroesophageal reflux disease and its complications. *Ann Intern Med*. 2005;143(3):199–211.
4. Kim J, Oh SW, Myung SK, et al. Association between coffee intake and gastroesophageal reflux disease: a *meta*-analysis. *Dis Esophagus*. 2014;27(4):311–317.
5. Katz PO, Gerson LB, Vela MF. Guidelines for the diagnosis and management of gastroesophageal reflux disease. *Am J Gastroenterol*. 2013;108(3):308–328. quiz 329.
6. Tosetti C, Savarino E, Benedetto E, De, Bastiani R. Study group for the evaluation of GERD triggering foods. Elimination of dietary triggers is successful in treating symptoms of gastroesophageal reflux disease. *Dig Dis Sci*. 2021;66(5):1565–1571.
7. Kaltenbach T, Crockett S, Gerson LB. Are lifestyle measures effective in patients with gastroesophageal reflux disease? an evidence-based approach. *Arch Intern Med*. 2006;166(9):965–971.
8. Newberry C, Lynch K. The role of diet in the development and management of gastroesophageal reflux disease: why we feel the burn. *J Thorac Dis*. 2019;11(suppl 12):S1594–S1601.

9. Fox M, Barr C, Nolan S, Lomer M, Anggiansah A, Wong T. The effects of dietary fat and calorie density on esophageal acid exposure and reflux symptoms. *Clin Gastroenterol Hepatol.* 2007;5(4):439–444.
10. Ruhl CE, Everhart JE. Overweight, but not high dietary fat intake, increases risk of gastroesophageal reflux disease hospitalization: the NHANES I epidemiologic followup study. first national health and nutrition examination survey. *Ann Epidemiol.* 1999;9(7):424–435.
11. Pointer SD, Rickstrew J, Slaughter JC, Vaezi MF, Silver HJ. Dietary carbohydrate intake, insulin resistance and gastro-oesophageal reflux disease: a pilot study in european- and african-american obese women. *Aliment Pharmacol Ther.* 2016;44(9):976–988.
12. Langella C, Naviglio D, Marino M, Calogero A, Gallo M. New food approaches to reduce and/or eliminate increased gastric acidity related to gastroesophageal pathologies. *Nutrition.* 2018;54:26–32.
13. Austin GL, Thiny MT, Westman EC, Yancy Jr WS, Shaheen NJ. A very low-carbohydrate diet improves gastroesophageal reflux and its symptoms. *Dig Dis Sci.* 2006;51(8):1307–1312.
14. Piesman M, Hwang I, Maydonovitch C, Wong RK. Nocturnal reflux episodes following the administration of a standardized meal. does timing matter? *Am J Gastroenterol.* 2007;102(10):2128–2134.
15. Fujiwara Y, Machida A, Watanabe Y, et al. Association between dinner-to-bed time and gastro-esophageal reflux disease. *Am J Gastroenterol.* 2005;100(12):2633–2636.
16. Jacobson BC, Somers SC, Fuchs CS, Kelly CP, Camargo Jr. CA. Body-mass index and symptoms of gastroesophageal reflux in women. *N Engl J Med.* 2006;354(22):2340–2348.
17. Park SK, Lee T, Yang HJ, et al. Weight loss and waist reduction is associated with improvement in gastroesophageal disease reflux symptoms: a longitudinal study of 15 295 subjects undergoing health checkups. *Neurogastroenterol Motil.* 2017;29(5). Available from: https://doi.org/10.1111/nmo.13009. Epub 2016 Dec 21.
18. de Bortoli N, Guidi G, Martinucci I, et al. Voluntary and controlled weight loss can reduce symptoms and proton pump inhibitor use and dosage in patients with gastroesophageal reflux disease: a comparative study. *Dis Esophagus.* 2016;29(2):197–204.
19. Singh S, Sharma AN, Murad MH, et al. Central adiposity is associated with increased risk of esophageal inflammation, metaplasia, and adenocarcinoma: a systematic review and *meta*-analysis. *Clin Gastroenterol Hepatol.* 2013;11(11):1399–1412. e7.
20. Chung TH, Lee J, Jeong ID, Lee KC. Effect of weight changes on the development of erosive esophagitis. *Korean J Fam Med.* 2020;41(1):14–19.
21. El-Serag H. The association between obesity and GERD: a review of the epidemiological evidence. *Dig Dis Sci.* 2008;53(9):2307–2312.
22. Nocon M, Labenz J, Jaspersen D, et al. Association of body mass index with heartburn, regurgitation and esophagitis: results of the progression of gastroesophageal reflux disease study. *J Gastroenterol Hepatol.* 2007;22(11):1728–1731.
23. Nilsson M, Johnsen R, Ye W, Hveem K, Lagergren J. Lifestyle related risk factors in the aetiology of gastro-oesophageal reflux. *Gut.* 2004;53(12):1730–1735.
24. Nandurkar S, Locke 3rd GR, Fett S, Zinsmeister AR, Cameron AJ, Talley NJ. Relationship between body mass index, diet, exercise and gastro-oesophageal reflux symptoms in a community. *Aliment Pharmacol Ther.* 2004;20(5):497–505.
25. Tramacere I, La Vecchia C, Negri E. Tobacco smoking and esophageal and gastric cardia adenocarcinoma: a *meta*-analysis. *Epidemiology.* 2011;22(3):344–349.
26. Andrici J, Cox MR, Eslick GD. Cigarette smoking and the risk of barrett's esophagus: a systematic review and *meta*-analysis. *J Gastroenterol Hepatol.* 2013;28(8):1258–1273.
27. Cook MB, Shaheen NJ, Anderson LA, et al. Cigarette smoking increases risk of barrett's esophagus: an analysis of the barrett's and esophageal adenocarcinoma consortium. *Gastroenterology.* 2012;142(4):744–753.

28. Villamil Morales IM, Gallego Ospina DM, Otero, Regino WA. Impact of head of bed elevation in symptoms of patients with gastroesophageal reflux disease: a randomized single-blind study (IBELGA). *Gastroenterol Hepatol.* 2020;43(6):310–321.
29. Khan BA, Sodhi JS, Zargar SA, et al. Effect of bed head elevation during sleep in symptomatic patients of nocturnal gastroesophageal reflux. *J Gastroenterol Hepatol.* 2012;27(6):1078–1082.
30. Fass R, Achem SR, Harding S, Mittal RK, Quigley E. Review article: supra-oesophageal manifestations of gastro-oesophageal reflux disease and the role of night-time gastro-oesophageal reflux. *Aliment Pharmacol Ther.* 2004;20(suppl 9):26–38.
31. Miyazaki H, Igarashi A, Takeuchi T, et al. Vonoprazan vs proton-pump inhibitors for healing gastroesophageal reflux disease: a systematic review. *J Gastroenterol Hepatol.* 2019;34(8):1316–1328.
32. Cossentino MJ, Mann K, Armbruster SP, Lake JM, Maydonovitch C, Wong RK. Randomised clinical trial: the effect of baclofen in patients with gastro-oesophageal reflux—a randomised prospective study. *Aliment Pharmacol Ther.* 2012;35(9):1036–1044.
33. Kahrilas PJ, McColl K, Fox M, et al. The acid pocket: a target for treatment in reflux disease? *Am J Gastroenterol.* 2013;108(7):1058–1064.
34. Ramirez B, Richter JE. Review article: promotility drugs in the treatment of gastro-oesophageal reflux disease. *Aliment Pharmacol Ther.* 1993;7(1):5–20.
35. Graham DY, Tansel A. Interchangeable use of proton pump inhibitors based on relative *potency. Clin Gastroenterol Hepatol.* 2018;16(6):800–808.e7.
36. Maret-Ouda J, Markar SR, Lagergren J. Gastroesophageal reflux disease: a review. *JAMA.* 2020;324(24):2536–2547.
37. Inadomi JM, Jamal R, Murata GH, et al. Step-down management of gastroesophageal reflux disease. *Gastroenterology.* 2001;121(5):1095–1100.
38. Gunaratnam NT, Jessup TP, Inadomi J, Lascewski DP. Sub-optimal proton pump inhibitor dosing is prevalent in patients with poorly controlled gastro-oesophageal reflux disease. *Aliment Pharmacol Ther.* 2006;23(10):1473–1477.
39. Wolfe MM, Sachs G. Acid suppression: optimizing therapy for gastroduodenal ulcer healing, gastroesophageal reflux disease, and stress-related erosive syndrome. *Gastroenterology.* 2000;118(2 suppl 1):S9–S31.
40. Kahrilas PJ, Shaheen NJ, Vaezi MF, et al. American gastroenterological association medical position statement on the management of gastroesophageal reflux disease. *Gastroenterology.* 2008;135(4):1383–1391. 1391.e1–5.
41. Wilder-Smith C, Röhss K, Bokelund Singh S, Sagar M, Nagy P. The effects of dose and timing of esomeprazole administration on 24-h, daytime and night-time acid inhibition in healthy volunteers. *Aliment Pharmacol Ther.* 2010;32(10):1249–1256.
42. Heidelbaugh JJ, Goldberg KL, Inadomi JM. Overutilization of proton pump inhibitors: a review of cost-effectiveness and risk corrected. *Am J Gastroenterol.* 2009;104 (suppl 2):S27–S32.
43. Shirai N, Furuta T, Moriyama Y, et al. Effects of CYP2C19 genotypic differences in the metabolism of omeprazole and rabeprazole on intragastric pH. *Aliment Pharmacol Ther.* 2001;15(12):1929–1937.
44. Hillman L, Yadlapati R, Thuluvath AJ, Berendsen MA, Pandolfino JE. A review of medical therapy for proton pump inhibitor nonresponsive gastroesophageal reflux disease. *Dis Esophagus.* 2017;30(9):1–15.
45. Ishizaki T, Horai Y. Review article: cytochrome P450 and the metabolism of proton pump inhibitors—emphasis on rabeprazole. *Aliment Pharmacol Ther.* 1999;13(suppl 3):27–36.
46. Yadlapati R, Vaezi MF, Vela MF, et al. Management options for patients with GERD and persistent symptoms on proton pump inhibitors: recommendations from an expert panel. *Am J Gastroenterol.* 2018;113(7):980–986.

47. Roman S, Mion F. Refractory GERD, beyond proton pump inhibitors. *Curr Opin Pharmacol.* 2018;43:99–103.
48. Vigneri S, Termini R, Leandro G, et al. A comparison of five maintenance therapies for reflux esophagitis. *N Engl J Med.* 1995;333(17):1106–1110.
49. El-Serag HB, Aguirre TV, Davis S, Kuebeler M, Bhattacharyya A, Sampliner RE. Proton pump inhibitors are associated with reduced incidence of dysplasia in barrett's esophagus. *Am J Gastroenterol.* 2004;99(10):1877–1883.
50. Singh S, Garg SK, Singh PP, Iyer PG, El-Serag HB. Acid-suppressive medications and risk of oesophageal adenocarcinoma in patients with barrett's oesophagus: a systematic review and *meta*-analysis. *Gut.* 2014;63(8):1229–1237.
51. Shaheen NJ, Falk GW, Iyer PG, Gerson LB. American college of gastroenterology. ACG clinical guideline: diagnosis and management of barrett's esophagus. *Am J Gastroenterol.* 2016;111(1):30–50. quiz 51.
52. Spechler SJ, Sharma P, Souza RF, Inadomi JM, Shaheen NJAmerican Gastroenterological Association. American gastroenterological association medical position statement on the management of barrett's esophagus. *Gastroenterology.* 2011;140(3):1084–1091.
53. Bresalier RS. Chemoprevention of barrett's esophagus and esophageal adenocarcinoma. *Dig Dis Sci.* 2018;63(8):2155–2162.
54. Kushnir VM, Sayuk GS, Gyawali CP. Abnormal GERD parameters on ambulatory pH monitoring predict therapeutic success in noncardiac chest pain. *Am J Gastroenterol.* 2010;105(5):1032–1038.
55. Chang AB, Lasserson TJ, Gaffney J, Connor FL, Garske LA. Gastro-oesophageal reflux treatment for prolonged non-specific cough in children and adults. *Cochrane Database Syst Rev.* 2011;2011(1):CD004823.
56. Gibson PG, Henry RL, Coughlan JL. Gastro-oesophageal reflux treatment for asthma in adults and children. *Cochrane Database Syst Rev.* 2003;2:CD001496. doi(2):CD001496.
57. Durazzo M, Lupi G, Cicerchia F, et al. Extra-esophageal presentation of gastroesophageal reflux disease: 2020 update. *J Clin Med.* 2020;9(8):2559. Available from: https://doi.org/10.3390/jcm9082559.
58. Moayyedi P, Eikelboom JW, Bosch J, et al. Safety of proton pump inhibitors based on a large, multi-year, randomized trial of patients receiving rivaroxaban or aspirin. *Gastroenterology.* 2019;157(3):682–691.e2.
59. Li CM, Zhernakova A, Engstrand L, Wijmenga C, Brusselaers N. Systematic review with *meta*-analysis: the risks of proton pump inhibitors during pregnancy. *Aliment Pharmacol Ther.* 2020;51(4):410–420.
60. Acar S, Keskin-Arslan E, Uysal N, Karadaş B, Kaplan YC. Letter: safety of proton pump inhibitors during pregnancy. *Aliment Pharmacol Ther.* 2020;52(4):739.
61. Gill SK, O'Brien L, Einarson TR, Koren G. The safety of proton pump inhibitors (PPIs) in pregnancy: a *meta*-analysis. *Am J Gastroenterol.* 2009;104(6):1541–1545. quiz 1540, 1546.
62. Nikfar S, Abdollahi M, Moretti ME, Magee LA, Koren G. Use of proton pump inhibitors during pregnancy and rates of major malformations: a *meta*-analysis. *Dig Dis Sci.* 2002;47(7):1526–1529.
63. Helgadottir H, Bjornsson ES. Problems associated with deprescribing of proton pump inhibitors. *Int J Mol Sci.* 2019;20(21):5469. Available from: https://doi.org/10.3390/ijms20215469.
64. Miner Jr PB, Allgood LD, Grender JM. Comparison of gastric pH with omeprazole magnesium 20.6 mg (prilosec OTC) o.m. famotidine 10 mg (pepcid AC) b.d. and famotidine 20 mg b.d. over 14 days of treatment. *Aliment Pharmacol Ther.* 2007;25(1):103–109.
65. Sigterman KE, van Pinxteren B, Bonis PA, Lau J, Numans ME. Short-term treatment with proton pump inhibitors, H2-receptor antagonists and prokinetics for gastro-oesophageal reflux disease-like symptoms and endoscopy negative reflux disease. *Cochrane Database Syst Rev.* 2013;2013(5):CD002095.

66. Xue S, Katz PO, Banerjee P, Tutuian R, Castell DO. Bedtime H2 blockers improve nocturnal gastric acid control in GERD patients on proton pump inhibitors. *Aliment Pharmacol Ther.* 2001;15(9):1351–1356.
67. Abdul-Hussein M, Freeman J, Castell D. Concomitant administration of a Histamine2 receptor antagonist and proton pump inhibitor enhances gastric acid suppression. *Pharmacotherapy.* 2015;35(12):1124–1129.
68. Chiba N, De Gara CJ, Wilkinson JM, Hunt RH. Speed of healing and symptom relief in grade II to IV gastroesophageal reflux disease: a *meta*-analysis. *Gastroenterology.* 1997;112(6):1798–1810.
69. Fackler WK, Ours TM, Vaezi MF, Richter JE. Long-term effect of H2RA therapy on nocturnal gastric acid breakthrough. *Gastroenterology.* 2002;122(3):625–632.
70. Mahase E. FDA recalls ranitidine medicines over potential cancer causing impurity. *BMJ.* 2019;367:l5832.
71. Katzka DA, Kahrilas PJ. Advances in the diagnosis and management of gastroesophageal reflux disease. *BMJ.* 2020;371:m3786.
72. Gyawali CP, Fass R. Management of gastroesophageal reflux disease. *Gastroenterology.* 2018;154(2):302–318.
73. Castell DO, Dalton CB, Becker D, Sinclair J, Castell JA. Alginic acid decreases postprandial upright gastroesophageal reflux. comparison with equal-strength antacid. *Dig Dis Sci.* 1992;37(4):589–593.
74. Iwakiri K, Kinoshita Y, Habu Y, et al. Evidence-based clinical practice guidelines for gastroesophageal reflux disease 2015. *J Gastroenterol.* 2016;51(8):751–767.
75. Leiman DA, Riff BP, Morgan S, et al. Alginate therapy is effective treatment for GERD symptoms: a systematic review and *meta*-analysis. *Dis Esophagus.* 2017;30(5):1–9.
76. Simon B, Ravelli GP, Goffin H. Sucralfate gel vs placebo in patients with non-erosive gastro-oesophageal reflux disease. *Aliment Pharmacol Ther.* 1996;10(3):441–446.

CHAPTER 7

Endoscopic GERD therapeutics in obesity

Raj Shah[1,2], Christopher C. Thompson[1,2] and Pichamol Jirapinyo[1,2]
[1]Division of Gastroenterology, Hepatology and Endoscopy, Brigham and Women's Hospital, Boston, MA, United States
[2]Harvard Medical School, Boston, MA, United States

Introduction

There is an increasing prevalence of both gastroesophageal reflux disease (GERD) and obesity.[1,2] For patients who fail medical therapy for GERD, surgical fundoplication, such as laparoscopic Nissen fundoplication, is traditionally offered.[3–5] Nevertheless, the efficacy of laparoscopic Nissen fundoplication appears limited in patients with obesity. Traditionally, for patients with obesity and concomitant GERD, Roux-en-Y gastric bypass has been the standard of care.[4–6] Nevertheless, only patients with a body mass index (BMI) of at least 35 kg/m^2 are eligible, and the majority of these eligible patients do not elect to undergo the surgery, leaving the majority of this patient population undertreated.[7] Over the past decades, several endoscopic antireflux procedures have been developed.[8] While most studies assessed their efficacy in patients without obesity, there has been an increased interest in assessing their outcomes in patients with obesity. Furthermore, with recent development in endoscopic bariatric and metabolic therapies (EBMTs), these procedures may be offered to induce weight loss with secondary benefits on reflux symptoms. This chapter summarizes the currently available endoscopic antireflux procedures as well as EBMTs, with a focus specifically on patients suffering from obesity and concomitant GERD.

GERD is a major healthcare burden in the United States (US) and worldwide. Specifically, its prevalence in the US ranges from 18.1% to 27.8% with an increasing incidence over the past decades.[1] In 2010 there were more than seven million ambulatory visits for GERD in a year.[9] One of the major risk factors for GERD is obesity. Specifically, patients with obesity have a 1.94 increased risk of having GERD compared to those with normal weight.[10] This is likely multifactorial with possible

Obesity and Esophageal Disorders
DOI: https://doi.org/10.1016/B978-0-323-98365-5.00009-7

causes being an increase in intraabdominal pressure, a decrease in lower esophageal sphincter (LES) function, and an increased risk of developing a hiatal hernia.[11–13]

Great efforts have been made to determine optimal treatment options for those who suffer from GERD and obesity. Although proton pump inhibitors (PPIs) are effective for the treatment of GERD, 10%–40% of patients fail to respond to the therapy.[14] Furthermore, in patients who suffer from both GERD and obesity, the response rate to PPI has been reported to be even lower.[6] On the other end of the treatment spectrum, several antireflux surgeries have been described and are available. Surgical fundoplication involves wrapping the fundus around the distal esophagus to strengthen the gastroesophageal junction (GEJ). The procedure may be performed as a full fundoplication (Nissen) or partial fundoplication from an anterior approach (Dor) or posterior approach (Toupet) to minimize dysphagia following the procedure.[15,16] Another surgical option includes magnetic sphincter augmentation (LINX, J&J Ethicon, Bridgewater, NJ), which involves the placement of a bracelet of magnetic beads to augment the strength of the LES.[8,16] While antireflux surgeries appear effective, they are associated with a 1.46 greater risk for serious adverse events (SAEs) compared to medical therapy.[17,18] Furthermore, for patients with obesity, the response rate to antireflux surgery has been reported to be lower compared to those with normal weight.[6,19–22] Specifically, patients with a BMI of greater than 35 kg/m^2 have a 4.8 time higher risk of failing Nissen fundoplication compared to those with a BMI of less than 35 kg/m^2.[19,23] Additionally, antireflux surgeries do not provide significant weight loss, leaving obesity, which is a major contributing factor of GERD, untreated.[5]

Bariatric surgery is effective at treating obesity with secondary beneficial effects on GERD.[24,25] Currently, the two most commonly performed bariatric surgeries are sleeve gastrectomy (SG) and Roux-en-Y gastric bypass (RYGB).[24,26] For SG, several studies have demonstrated improvement in GERD symptoms following the procedure likely as a secondary effect of weight loss.[27–29] Nevertheless, SG has also been shown to be associated with worsening GERD symptoms or de novo GERD.[26,28,30–33] Genco et al. demonstrated an increase in the prevalence of GERD symptoms from 33.6% to 68.1% at 58 months following SG. Additionally, there was a 17.2% rate of a new diagnosis of nondysplastic Barrett's esophagus.[30] A recent systematic review and metaanalysis further confirmed this finding showing an 11.6% rate of Barrett's esophagus following SG even in the absence of presurgical GERD symptoms.[34] In contrast, RYGB has been shown to improve reflux symptoms, esophagitis, and incidence of GERD.[5] Specifically, a systematic review and metaanalysis showed greater improvement of GERD symptoms with RYGB compared to SG.[35] While data on the effect of bariatric surgery on GERD are promising, the surgery is limited to those with a BMI of at least 40 or 35 kg/m^2 with at least one obesity-related comorbidity. Furthermore, within those who are eligible for bariatric surgery, less than 2% elect to undergo the surgery, leaving the majority of patients with obesity and GERD undertreated.[7,24]

For the past decades, several endoscopic antireflux procedures, as well as EBMTs, have been developed as an alternative minimally-invasive approach to treat patients with GERD and obesity, respectively. In this chapter, we will discuss these procedures with a focus on their techniques, safety, and efficacy, specifically in the patient population with obesity.

Endoscopic antireflux procedures

Endoscopic antireflux procedures focus on decreasing the LES compliance and/or lengthening the intraabdominal segment of the esophagus.[36,37] Currently, there are several endoscopic devices that are specifically designed for GERD therapy.[8]

Radiofrequency therapy (Stretta)

Stretta therapy (Restech; Mederi Therapeutics, Greenwich, CT) utilizes radiofrequency energy to target the muscle fibers of the LES and cardia.[8,38] The Stretta system consists of the radiofrequency generator and the catheter system. The catheter is introduced over a guidewire and positioned sequentially at three levels: 0.5 cm proximal to the GEJ, at the GEJ, and 0.5 cm below the GEJ. At each level, the balloon is inflated and then four-needle tips (22-gague) are extended into the muscular layer to deliver a radiofrequency current. Subsequently, the catheter is rotated 45 degrees clockwise to deliver radiofrequency energy to four additional points. It is thought that the thermal injury helps increase LES muscle thickness, leading to a reduction in the frequency of transient LES relaxation (TLESR) and GEJ compliance while increasing LES pressure.[38,39]

To date, there have been several randomized controlled trials (RCTs) assessing the safety and efficacy of Stretta in comparison to a sham procedure or PPI. A recent metaanalysis including 24 observational and 4 RCT studies with 2468 patients demonstrated an improvement in a standardized heartburn score, incidence of erosive esophagitis, esophageal acid exposure time, and health-related quality of life (HRQL) following Stretta. Additionally, it showed that greater than half of the patients were able to discontinue PPI at follow-up after the Stretta procedure. The pooled adverse event (AE) rate was 0.93% including mucosal lacerations and erosions.[8,40]

In regards to the impact of obesity and response to Stretta, an abstract reported 10-year prospective outcomes of Stretta in 98 patients stratified by BMI class. While significant improvements in GERD-related outcomes were found for each of the subpopulations (BMI 18.5–24.9, 25–29.9 and ≥30.0), no difference among each BMI category was seen for GERD-HRQL and satisfaction.[41]

Transoral incisionless fundoplication

Transoral incisionless fundoplication (TIF) is performed using a fundoplication device (EsophyX, EndoGastric Solutions, Redmond, WA). The device allows the passage of an endoscope through its channel, both of which are then gently introduced into the stomach under visualization. The endoscope along the device is retroflexed, and a helix is used to grasp the tissue slightly distal to the Z line. The fundus of the stomach is then folded around the distal esophagus using the tissue mold. Suctioning is applied before deployment of H-shaped fasteners through apposed layers of esophageal and fundic tissue. This step is repeated to create a full-thickness, 270-degree circumferential wrap and a 2–4-cm long gastroesophageal flap valve.[42] Eligibility criteria for TIF includes symptomatic chronic GERD in patients who require and respond to pharmacotherapy and hiatal hernia ≤2 cm and BMI less than 35 kg/m^2.[8] Most studies also excluded patients with LA Grade C or D esophagitis; therefore TIF is usually avoided in these patients.[8] Additionally, patients with Hill grade I or II appear to respond better to TIF compared to those with III and IV.[43]

A systematic review and metaanalysis of 32 observational and RCT studies with 1475 patients showed significant improvements in GERD symptom score (mean difference (MD) of 23.78), reflux symptom index (RSI) (MD of 14.28), DeMeester score (MD of 10.22), and GERD-HRQL (MD of 17.72) following TIF. Additionally, 89% of patients were able to discontinue PPI following the procedure.[44] Another systematic review and metaanalysis of 16 studies with 781 patients showed a pooled

SAE rate of 2.4%, which included perforation (0.90%) bleeding (0.64%), pneumothorax (0.51%), and severe epigastric pain (0.13%).[45] The TEMPO open-label trial, comparing TIF and maximal dose PPI, demonstrated a 90% rate of PPI cessation at 6 months following TIF.[46] At 5 years, the proportion of patients who remained off PPI was 46%, demonstrating the durability of the procedure.[47] Of note, the average BMI of the patients included in the 5-year study was 28.5 ± 3.7 kg/m^2. Therefore durability data for TIF in patients with BMI greater than 30 remains limited.[47] A network metaanalysis of 7 RCTs comparing TIF, Nissen fundoplication, PPI, and sham procedure showed that TIF was associated with the most improvement in HRQL, while Nissen fundoplication was associated with the most increase in percent time at pH <4 and LES pressure.[48] Additionally, a randomized sham-controlled trial (RESPECT trial) of 129 patients assessed TIF versus omeprazole on regurgitation symptoms. The study showed the superiority of TIF over PPI to eliminate regurgitation (67% vs 45%) at 6 months. During this trial, the mean operating time was 49 minutes, and the mean number of fasteners used was 23. In regards to patients with concomitant obesity, 51.7% of patients who underwent TIF in this study had a BMI of 25–30 kg/m^2 and 23% had a BMI greater than 30 kg/m^2, but less than 35 kg/m^2.[49] Specific outcome data in patients with obesity stratified by BMI remains scarce.

For a subgroup of patients who have a large hiatal hernia of greater than 2 cm or Hill grade III and IV, a combination of laparoscopic hiatal hernia repair followed by TIF during the same session has been proposed as a treatment option. This procedure has been termed as concomitant transoral incisionless fundoplication, also known as c-TIF. An observational study of 60 patients who underwent c-TIF showed 100% technical success and a marked reduction in both DeMeester score (43.7 to 4.9) and acid exposure time (12.7 to 1.28%). Moreover, significant improvements in reflux questionnaire scores for symptom frequency and severity at 6 months, GERD-HRQL index scores for heartburn and regurgitation, and RSI scores were also found. Baseline demographics showed a mean BMI of 30 kg/m^2 with a range of 19.8–36 kg/m^2. Stratified outcomes based on BMI, however, were not reported.[50]

Overall, TIF is an alternative to medical and surgical treatments for GERD. For patients with obesity, TIF appears effective in patients with class I obesity with limited data in patients with a BMI of greater than 35 kg/m^2.

Endoscopic antireflux procedures in patients with altered anatomy

For patients who have undergone SG or RYGB for weight loss, their GERD symptoms may persist and/or de novo GERD symptoms may develop.[25,31] Traditionally, this may be treated with the conversion of SG to RYGB or surgical revision of RYGB. Nevertheless, these surgeries are usually associated with a higher morbidity rate than the index procedure.[51–54] Alternatively, endoscopic revisions of SG and RYGB may be offered to induce weight loss, which subsequently may result in improvement in GERD symptoms.[55,56] Furthermore, several endoscopic techniques specifically designed to strengthen the GEJ using an antegrade approach (rather than a retroflexed approach) have been described.[57]

The mucosal ablation and suturing at the esophagogastric junction (MASE) procedure consists of argon plasma coagulation below the GEJ to improve tissue approximation, followed by endoscopic suturing in an antegrade view along the lesser curvature of the cardia to mitigate reflux. A case series of 27 patients, 7 (26%) of whom had altered anatomy, demonstrated that 57% of the patients were able to discontinue or reduce the dose of PPI following the procedure. Average procedure time was 25 minutes. The most common AE was self-limited epigastric pain (22%).[58]

The resection and plication procedure combines endoscopic mucosal resection (EMR) and endoscopic suturing to tighten the GEJ. This was recently shown to be technically feasible with 80% of the patients being able to discontinue PPI following the procedure.[57,59] Additionally, this procedure can be performed in a single session along with an endoscopic revision of gastrojejunal anastomosis in patients with a RYGB anatomy who struggle from weight regain as well as GERD.[57]

The antireflux mucosectomy (ARM) procedure refers to when EMR or endoscopic submucosal dissection is performed in a retroflexed manner to induce scar formation and therefore tighten the GEJ. In a pilot study of 10 patients, ARM was associated with a significant reduction in DeMeester score, mean heartburn score, regurgitation score, and total score. Out of the 10 patients, 2 underwent total circumference resection, both of whom experienced stenosis, which was treated successfully with endoscopic dilation.[57,60] This technique has been described to be technically feasible and effective for the treatment of GERD in patients with SG anatomy.[61]

Endoscopic weight loss procedures

Weight loss in patients who are overweight or have obesity is correlated with improvement in GERD symptoms. Specifically, it has been shown that a reduction of 5–10 cm of waist circumference or 5%–10% of baseline weight in women and ≥ 10 cm of waist circumference or ≥ 10% of baseline weight in men resulted in a significant reduction of overall GERD scores.[62] To date, there are several EBMTs that are available in the US that are associated with at least 5% or 10% total body weight loss. These include intragastric balloons (IGB), endoscopic sleeve gastroplasty (ESG), and aspiration therapy.

Intragastric balloon

While there are several IGB that are approved by the Food and Drug Administration (FDA), two are currently available commercially—Orbera (Apollo Endosurgery, Austin, TX) and Obalon (ReShape Sciences, San Clemente, CA). IGBs are a therapeutic weight loss option for patients with a BMI of 30–40 kg/m^2.[63,64] Orbera is a silicone-based spherical balloon that is placed and removed via endoscopy. It is filled with saline, approximately 500–600 cc, using a detachable catheter and remains in the stomach for 6 months.[65] Alternatively, the Obalon system is a swallowable gas-filled balloon that is removed via endoscopy at 6 months. The system comprises of three nylon polyethylene balloons, each filled with 250 cc of nitrogen-mix gas using a 3 Fr catheter, swallowed between 3 and 9 weeks of each other.[66]

Recently, the American Gastroenterological Association 2021 guideline suggested the use of IGBs in patients with obesity seeking a weight loss intervention based on moderate certainty of evidence. In this guideline, it has been demonstrated that a higher proportion of patients who underwent IGB therapy with concomitant lifestyle modification were able to achieve at least 5% and 10% weight loss at 6 and 12 months compared to lifestyle modification alone.[63,64,67] Additionally, a metaanalysis conducted by the American Society of Gastrointestinal Endoscopy Bariatric Endoscopy Taskforce reported a mean of 14.4% total body weight loss at 12 months and 9.7% at 36 months with the Orbera balloon.[68]

While IGB is effective at inducing weight loss, its effect on GERD remains controversial. It has been hypothesized that IGB may increase intragastric pressure, which may increase the incidence of GERD.[69] Hirsch et al. showed an increase in GERD symptoms as well as an increase in

TLESR up to 20 weeks following IGB placement.[70] Another observation study by Barrichello et al. showed an increase in DeMeester score following IGB placement, although no significant changes in LES pressure were observed.[71] Furthermore, a systematic review and metaanalysis focused on the Orbera balloon suggested a filling volume of 600–650 cc as higher balloon volumes were found to be associated with a lower rate of esophagitis.[72,73] Nevertheless, another study by Tolone et al. demonstrated no significant difference in LES pressure, frequency in normal and ineffective peristalsis, total acid exposure, and total number of reflux detected using impedance pH monitoring before and after IGB placement.[74] Therefore the association between GERD and IGB requires further study.

Endoscopic sleeve gastroplasty

ESG involves the use of an endoscopic suturing device (Overstitch, Apollo Endosurgery) or endoscopic plication device (Incisionless Operating Platform, USGI Medical, San Clemente, CA) to reduce the gastric volume along the greater curvature. Both devices have FDA clearance for tissue apposition.[75] ESG is typically offered to patients with a BMI between 30 and 40 kg/m^2 or higher than 40 kg/m^2 if they are not a surgical candidate or do not want to undergo surgery.[76]

ESG via a suturing technique involves the use of the suturing device to deliver full-thickness bites in the gastric body to reduce its volume.[77] The device consists of a curved suture arm, needle pick-up, needle that is attached to a 2/0 suture, and a tissue helix.[75,78] During the procedure, the tissue helix is used to grasp and bring the tissue into the jaw of the suturing device to ensure full-thickness bites. While several suturing patterns have been described, most start with placing the first suture in the distal gastric body, then moving proximally to end the last suture at the junction between the gastric body and fundus.[79–81]

A systematic review and metaanalysis of eight observational studies including 1772 patients showed 16.5% total body weight loss at 12 months. A pooled serious adverse event rate was 2.2% including nausea or pain requiring hospitalization, upper GI bleeding, perigastric leak or collection, pulmonary embolism, and pneumoperitoneum.[82]

Another approach to perform ESG is via a plication technique, a procedure also known as primary obesity surgery endoluminal (POSE). The procedure is performed using an endoscopic plication device, which consists of a 54-F flexible transport with four working channels, g-Lix for grasping tissue, g-Prox for approximating tissue, and g-Caths that consist of a hollow needle with a pair of snowshoe suture anchors for tissue plication. The original POSE procedure focuses on placing plications in the fundus with the goal of reducing gastric accommodation. More recently, a modified technique known as distal POSE or POSE 2.0, has been described where plications are placed primarily in the gastric body with the goal of altering gastric motility. During this procedure, plications are placed along the width and length of the gastric body to narrow and shorten the stomach. Additionally, following this technique, the intraabdominal esophagus has been shown to be elongated, leading to lengthening of the high-pressure zone of the esophagus. This has been postulated to have a beneficial impact on GERD symptoms following the procedure.[83,84]

A systematic review and metaanalysis of seven studies with 613 patients who underwent the original POSE procedure demonstrated a 12.68% total body weight loss at 12–15 months. A pooled SAE rate was 2.84% and included extraluminal bleeding, GI bleeding, hepatic abscess, and severe pain/nausea/vomiting.[85] Regarding the distal POSE technique, a case series of 10 patients reported the efficacy of the procedure on weight loss and GERD symptoms. The average weight loss was 15% total body weight loss at 6 months. Out of 10, 3 patients had underlying GERD. Following the procedure, one patient was able to discontinue PPI and two were able to decrease PPI dosage.[83]

Compared to its surgical correlate (SG), ESG is thought to be associated with a lower incidence of GERD following the procedure. A case-control study comparing ESG to SG found a lower rate of new-onset GERD following ESG compared to SG (1.9% vs 14.5%).[86] This observation is thought to be due to several reasons including preservation of the angle of His and sling fibers of the LES, as well as lower intragastric pressure following ESG compared to SG.[87,88]

Another emerging therapy for the treatment of GERD in patients with obesity is a combination of the endoscopic antireflux procedure with EBMT. Specifically, a combination of TIF and ESG during the same session was previously described in a patient with obesity and concomitant GERD with a baseline BMI of 36 kg/m^2 on twice a day dosing PPI. TIF followed by ESG was successfully performed during the same session. At 2 months, the patient experienced 18% total body weight loss with no reflux symptoms on a daily dosing PPI.[89]

Similarly, a combination of endoscopic antireflux and endoscopic weight loss procedures can also be performed using a plication technique. In another case report, a patient with class I obesity and GERD underwent the endoscopic gastric plication procedure to treat GERD and obesity in a single session. In the retroflexed scope position, plications are placed at the GEJ 270 degrees around the cardia to tighten the GEJ and elongate the esophagus. Subsequently, the plication device is unretroflexed and the distal POSE procedure is completed with additional plications being placed in the antegrade fashion in the gastric body. At 6 months, a patient experienced 15.8% total body weight loss, 5-point BMI reduction and had resolution of GERD symptoms off PPI.[36]

Overall, ESG via suturing or plication is effective at treating obesity with the majority of patients achieving the weight loss threshold required for improvement in GERD symptoms. Additionally, the procedure appears to be physiologically favorable, especially for those with obesity and underlying GERD as it preserves the integrity of the LES and maintains a low intraluminal pressure, in contrast to its surgical analog: SG. Novel techniques combining ESG with other endoscopic antireflux procedures, such as TIF or endoscopic plication of the GEJ, appears feasible and may be performed in a single session for this patient population.

Aspiration therapy

Aspiration therapy (AspireAssist; Aspire Bariatric, King of Prussia, PA) is approved by the FDA for patients with a BMI of 35–55 kg/m^2. The device consists of a 26 Fr percutaneous endoscopic gastrostomy tube (A-tube) that is placed endoscopically via a standard pull technique. Approximately 2 weeks following A-tube placement, the tube is fitted to a Skin-Port. A Connector attaches to the Skin-Port to allow for the opening of the Skin-Port valve to remove about 30% of ingested calories about 30 minutes after a meal.[90,91] The US pivotal study (PATHWAY trial) assessing aspiration therapy demonstrated 14.2% total body weight loss in the intervention group versus 4.9% in the lifestyle modification alone group. The SAE rate was 3.6% and included severe abdominal pain, peritonitis, prepyloric ulcer, and skin-port malfunction.[91] A systematic review and metaanalysis showed that the amount of weight loss was maintained at 18.6% at 4 years following aspiration therapy.[92] To date, the data specifically on the efficacy of aspiration therapy on GERD symptoms remain limited.[69]

Summary

Endoscopic GERD therapy in patients with obesity is a rapidly evolving field. Improvement in GERD symptoms can be achieved by either

Table 7.1 Endoscopic therapeutic options for patients with GERD and obesity that are available and under study.

Endoscopic antireflux	Endoscopic weight loss	Combination therapies
Transoral incisionless fundoplication (TIF) (retroflexion)	Aspiration therapy	Endoscopic sleeve gastroplasty (ESG) + TIF
Plication (retroflexion)	Space-occupying devices	Endoscopic sleeve gastroplasty (POSE) + plication
Radiofrequency therapy (antegrade)	Endoscopic sleeve gastroplasty (ESG) via suturing or plication	
Mucosal ablation and suturing (MASE) (antegrade)		
Resection and plication (RAP) (antegrade)		
Antireflux mucosectomy (ARM) (retroflexion)		

anatomic correction of the gastroesophageal area or through weight loss itself. Interventions targeting both of these mechanisms of action appear to be promising. Further high-quality studies elucidating the optimal techniques, combinations of therapies, outcomes for patients with obesity, patient selection criteria, and endoscopic interventions for patients with a BMI above 35 are needed. Several endoscopic interventions for the treatment of GERD and obesity are currently available and can provide relief and clinical improvement to patients (Table 7.1). Clinicians should understand patients' values and preferences and counsel patients on the importance of treating both GERD and obesity and different treatment options.

Conflicts of interest

Raj Shah has no conflicts to disclose.

Christopher C. Thompson has the following disclosures: Apollo Endosurgery—consultant/research support (consulting fees/institutional research grants), Aspire Bariatrics—research support (institutional research grant), BlueFlame Healthcare Venture Fund—general partner, Boston Scientific—consultant (consulting fees), Covidien/Medtronic—consultant (consulting fees), EnVision Endoscopy (board member), Fractyl—consultant/advisory board member (consulting fees), GI Dynamics—consultant (consulting fees)/research support (institutional research grant), GI Windows—ownership interest, Olympus/Spiration—consultant (consulting fees)/research support (equipment loans), Spatz—research support (institutional research grant), USGI Medical—consultant (consulting fees)/ advisory board member (consulting fees)/research support (research grant).

Pichamol Jirapinyo: Apollo EndoSurgery—research support, Boston Scientific—research support, Endogastric Solutions—consultant, ERBE—consultant, Fractyl—research support, GI Dynamics—consultant, research support, Spatz—consultant, USGI Medical—research support.

Funding

None.

Author contributions

All authors approved the final submission.

References

1. El-Serag HB, Sweet S, Winchester CC, et al. Update on the epidemiology of gastro-oesophageal reflux disease: a systematic review. *Gut*. 2014;63:871–880.
2. Ward ZJ, Bleich SN, Cradock AL, et al. Projected US state-level prevalence of adult obesity and severe obesity. *N Engl J Med*. 2019;381:2440–2450.
3. Stefanidis D, Hope WW, Kohn GP, et al. Guidelines for surgical treatment of gastroesophageal reflux disease. *Surg Endosc*. 2010;24:2647–2669.
4. Katz PO, Dunbar KB, Schnoll-Sussman FH, et al. ACG clinical guideline for the diagnosis and management of gastroesophageal reflux disease. *Am J Gastroenterol*. 2022;117:27–56.
5. Park S, Weg R, Enslin S, et al. Ten things every gastroenterologist should know about antireflux surgery. *Clin Gastroenterol Hepatol*. 2020;18:1923–1929.
6. Khan A, Kim A, Sanossian C, et al. Impact of obesity treatment on gastroesophageal reflux disease. *World J Gastroenterol*. 2016;22:1627–1638.
7. English WJ, DeMaria EJ, Hutter MM, et al. American society for metabolic and bariatric surgery 2018 estimate of metabolic and bariatric procedures performed in the United States. *Surg Obes Relat Dis*. 2020;16:457–463.
8. Vaezi MF, Shaheen NJ, Muthusamy VR. State of evidence in minimally invasive management of gastroesophageal reflux: findings of a scoping review. *Gastroenterology*. 2020;159:1504–1525.
9. Peery AF, Crockett SD, Barritt AS, et al. Burden of gastrointestinal, liver, and pancreatic diseases in the United States. *Gastroenterology*. 2015;149:1731–1741. e3.

10. Acosta A, Streett S, Kroh MD, et al. White paper AGA: POWER – practice guide on obesity and weight management, education, and resources. *Clin Gastroenterol Hepatol*. 2017;15:631–649. e10.
11. Koo JS, Lee SW, Park SM, et al. Abdominal obesity as a risk factor for the development of erosive esophagitis in subjects with a normal esophago-gastric junction. *Gut Liver*. 2009;3:276–284.
12. Vaezi MF. GERD and obesity: a real BIG issue!. *Gastroenterology*. 2008;134:882–883.
13. Chang P, Friedenberg F. Obesity and GERD. *Gastroenterol Clin North Am*. 2014;43:161–173.
14. Yadlapati R, Vaezi MF, Vela MF, et al. Management options for patients with GERD and persistent symptoms on proton pump inhibitors: recommendations from an expert panel. *Am J Gastroenterol*. 2018;113:980–986.
15. Coletta D, Del Basso C, Ossola P, et al. Comparison between robotic Nissen, Toupet and Dor fundoplication: results. *Laparosc Surg*. 2021;5. 10-10.
16. Yadlapati R, Hungness ES, Pandolfino JE. Complications of antireflux surgery. *Am J Gastroenterol*. 2018;113:1137–1147.
17. Spechler SJ, Hunter JG, Jones KM, et al. Randomized trial of medical vs surgical treatment for refractory heartburn. *N Engl J Med*. 2019;381:1513–1523.
18. Garg SK, Gurusamy KS. Laparoscopic fundoplication surgery vs medical management for gastro-oesophageal reflux disease (GORD) in adults. *Cochrane Database Syst Rev*. 2015;CD003243.
19. Fontan FM, Carroll RS, Thompson D, et al. Current management of gastroesophageal reflux disease in the obese population – a review of the literature. *Mini-invasive Surg*. 2020;2020.
20. Prachand VN, Alverdy JC. Gastroesophageal reflux disease and severe obesity: fundoplication or bariatric surgery? *World J Gastroenterol*. 2010;16:3757–3761.
21. Schietroma M, Piccione F, Clementi M, et al. Short- and long-term, 11–22 years, results after laparoscopic Nissen Fundoplication in obese vs nonobese patients. *J Obes*. 2017;2017:7589408.
22. Schizas D, Mastoraki A, Papoutsi E, et al. LINX® reflux management system to bridge the "treatment gap" in gastroesophageal reflux disease: a systematic review of 35 studies. *World J Clin Cases*. 2020;8:294–305.
23. Morgenthal CB, Lin E, Shane MD, et al. Who will fail laparoscopic Nissen fundoplication? Preoperative prediction of long-term outcomes. *Surg Endosc*. 2007;21:1978–1984.
24. Jirapinyo P, Thompson CC. Obesity primer for the practicing gastroenterologist. *Am J Gastroenterol*. 2021;116:918–934.
25. Frezza EE, Ikramuddin S, Gourash W, et al. Symptomatic improvement in gastroesophageal reflux disease (GERD) following laparoscopic Roux-en-Y gastric bypass. *Surg Endosc*. 2002;16:1027–1031.
26. Ozsoy Z, Demir E. Which bariatric procedure is the most popular in the world? A bibliometric comparison. *Obes Surg*. 2018;28:2339–2352.
27. Melissas J, Daskalakis M, Koukouraki S, et al. Sleeve gastrectomy-a "food limiting" operation. *Obes Surg*. 2008;18:1251–1256.
28. Stenard F, Iannelli A. Laparoscopic sleeve gastrectomy and gastroesophageal reflux. *World J Gastroenterol*. 2015;21:10348–10357.
29. Daes J, Jimenez ME, Said N, et al. Laparoscopic sleeve gastrectomy: symptoms of gastroesophageal reflux can be reduced by changes in surgical technique. *Obes Surg*. 2012;22:1874–1879.
30. Genco A, Soricelli E, Casella G, et al. Gastroesophageal reflux disease and Barrett's esophagus after laparoscopic sleeve gastrectomy: a possible, underestimated long-term complication. *Surg Obes Relat Dis*. 2017;13:568–574.

31. DuPree CE, Blair K, Steele SR, et al. Laparoscopic sleeve gastrectomy in patients with preexisting gastroesophageal reflux disease: a national analysis. *JAMA Surg*. 2014; 149:328–334.
32. McCarty TR, Thompson CC. How to manage sleeve complications through endoscopy: gastroesophageal reflux disease. In: Al-Sabah S, Aminian A, Angrisani L, Haddad EA, Kow L, eds. *Laparoscopic Sleeve Gastrectomy*. Springer; 2021:499–506.
33. McCarty TR, Thompson CC. How to manage sleeve complications through endoscopy: strictures. In: Al-Sabah S, Aminian A, Angrisani L, Haddad EA, Kow L, eds. *Laparoscopic Sleeve Gastrectomy*. Springer; 2021:477–489.
34. Qumseya BJ, Qumsiyeh Y, Ponniah SA, et al. Barrett's esophagus after sleeve gastrectomy: a systematic review and meta-analysis. *Gastrointest Endosc*. 2021;93:343–352. e2.
35. Gu L, Chen B, Du N, et al. Relationship between bariatric surgery and gastroesophageal reflux disease: a systematic review and meta-analysis. *Obes Surg*. 2019;29: 4105–4113.
36. Jirapinyo P, Thompson CC. Endoscopic gastric plication for the treatment of GERD and underlying class I obesity. *VideoGIE*. 2021;6:74–76.
37. Committee AT, Thosani N, Goodman A, et al. Endoscopic anti-reflux devices (with videos). *Gastrointest Endosc*. 2017;86:931–948.
38. Triadafilopoulos G. Stretta: a valuable endoscopic treatment modality for gastroesophageal reflux disease. *World J Gastroenterol*. 2014;20:7730–7738.
39. Dughera L, Rotondano G, De Cento M, et al. Durability of stretta radiofrequency treatment for GERD: results of an 8-year follow-up. *Gastroenterol Res Pract*. 2014;2014:531907.
40. Fass R, Cahn F, Scotti DJ, et al. Systematic review and meta-analysis of controlled and prospective cohort efficacy studies of endoscopic radiofrequency for treatment of gastroesophageal reflux disease. *Surg Endosc*. 2017;31:4865–4882.
41. Noar MD, Squires P, Kahn S. Tu1228 sphincter targeted endoscopic therapy for GERD demonstrates obesity is not a factor in response to endoluminal reflux therapy. *Gastrointest Endosc*. 2017;85.
42. Ihde GM. The evolution of TIF: transoral incisionless fundoplication. *Ther Adv Gastroenterol*. 2020;13. 1756284820924206.
43. Sami Trad K. Transoral incisionless fundoplication: current status. *Curr Opin Gastroenterol*. 2016;32:338–343.
44. McCarty TR, Itidiare M, Njei B, et al. Efficacy of transoral incisionless fundoplication for refractory gastroesophageal reflux disease: a systematic review and meta-analysis. *Endoscopy*. 2018;50:708–725.
45. Huang X, Chen S, Zhao H, et al. Efficacy of transoral incisionless fundoplication (TIF) for the treatment of GERD: a systematic review with meta-analysis. *Surg Endosc*. 2017;31:1032–1044.
46. Trad KS, Barnes WE, Simoni G, et al. Transoral incisionless fundoplication effective in eliminating GERD symptoms in partial responders to proton pump inhibitor therapy at 6 months: the TEMPO randomized clinical trial. *Surg Innov*. 2015;22:26–40.
47. Trad KS, Barnes WE, Prevou ER, et al. The TEMPO trial at 5 years: transoral fundoplication (TIF 2.0) is safe, durable, and cost-effective. *Surg Innov*. 2018;25:149–157.
48. Richter JE, Kumar A, Lipka S, et al. Efficacy of laparoscopic Nissen Fundoplication vs transoral incisionless fundoplication or proton pump inhibitors in patients with gastroesophageal feflux disease: a systematic review and network meta-analysis. *Gastroenterology*. 2018;154:1298–1308. e7.
49. Hunter JG, Kahrilas PJ, Bell RC, et al. Efficacy of transoral fundoplication vs omeprazole for treatment of regurgitation in a randomized controlled trial. *Gastroenterology*. 2015;148:324–333. e5.

50. Choi AY, Roccato MK, Samarasena JB, et al. Novel interdisciplinary approach to GERD: concomitant laparoscopic hiatal hernia repair with transoral incisionless fundoplication. *J Am Coll Surg*. 2021;232:309–318.
51. Landreneau JP, Strong AT, Rodriguez JH, et al. Conversion of sleeve gastrectomy to Roux-en-Y gastric bypass. *Obes Surg*. 2018;28:3843–3850.
52. Qiu J, Lundberg PW, Javier Birriel T, et al. Revisional bariatric surgery for weight regain and refractory complications in a single MBSAQIP accredited center: what are we dealing with? *Obes Surg*. 2018;28:2789–2795.
53. Chaar ME, Lundberg P, Stoltzfus J. Thirty-day outcomes of sleeve gastrectomy vs Roux-en-Y gastric bypass: first report based on metabolic and bariatric surgery accreditation and quality improvement program database. *Surg Obes Relat Dis*. 2018;14:545–551.
54. Morales MP, Wheeler AA, Ramaswamy A, et al. Laparoscopic revisional surgery after Roux-en-Y gastric bypass and sleeve gastrectomy. *Surg Obes Relat Dis*. 2010;6:485–490.
55. Jirapinyo P, Kumar N, AlSamman MA, et al. Five-year outcomes of transoral outlet reduction for the treatment of weight regain after Roux-en-Y gastric bypass. *Gastrointest Endosc*. 2020;91:1067–1073.
56. Jirapinyo P, de Moura DTH, Thompson CC. Sleeve in sleeve: endoscopic revision for weight regain after sleeve gastrectomy. *VideoGIE*. 2019;4:454–457.
57. Chang KJ. Endoscopic foregut surgery and interventions: The future is now. The state-of-the-art and my personal journey. *World J Gastroenterol*. 2019;25:1–41.
58. Fortinsky KJ, Shimizu T, Chin MA, et al. Tu1168 Mucosal ablation and suturing at the esophagogastric junction (mase): a novel procedure for the management of patients with gastroesophageal reflux disease. *Gastrointest Endosc*. 2018;87.
59. Benias PC, D'Souza L, Lan G, et al. Initial experience with a novel resection and plication (RAP) method for acid reflux: a pilot study. *Endosc Int Open*. 2018;6: E443–E449.
60. Inoue H, Ito H, Ikeda H, et al. Anti-reflux mucosectomy for gastroesophageal reflux disease in the absence of hiatus hernia: a pilot study. *Ann Gastroenterol*. 2014;@ 27:346–351.
61. Hathorn KE, Jirapinyo P, Thompson CC. Endoscopic management of gastroesophageal reflux disease after sleeve gastrectomy by use of the antireflux mucosectomy procedure. *VideoGIE*. 2019;4:251–253.
62. Singh M, Lee J, Gupta N, et al. Weight loss can lead to resolution of gastroesophageal reflux disease symptoms: a prospective intervention trial. *Obes (Silver Spring)*. 2013;21:284–290.
63. Muniraj T, Day LW, Teigen LM, et al. AGA clinical practice guidelines on intragastric balloons in the management of obesity. *Gastroenterology*. 2021;160:1799–1808.
64. Shah R, Davitkov P, Abu Dayyeh BK, et al. AGA technical review on intragastric balloons in the management of obesity. *Gastroenterology*. 2021;160:1811–1830.
65. Courcoulas A, Abu Dayyeh BK, Eaton L, et al. Intragastric balloon as an adjunct to lifestyle intervention: a randomized controlled trial. *Int J Obes (Lond)*. 2017;41:427–433.
66. Sullivan S, Swain J, Woodman G, et al. Randomized sham-controlled trial of the 6-month swallowable gas-filled intragastric balloon system for weight loss. *Surg Obes Relat Dis*. 2018;14:1876–1889.
67. Shah R, Davitkov P, Abu Dayyeh BK, et al. Spotlight: intragastric balloons in the management of obesity. *Gastroenterology*. 2021;160:1810.
68. Force ABET, Committee ATAbu Dayyeh BK, et al. ASGE bariatric endoscopy task force systematic review and meta-analysis assessing the ASGE PIVI thresholds for adopting endoscopic bariatric therapies. *Gastrointest Endosc*. 2015;82:425–438. e5.

69. Fass OZ, Mashimo H. The effect of bariatric surgery and endoscopic procedures on gastroesophageal reflux disease. *J Neurogastroenterol Motil*. 2021;27:35–45.
70. Hirsch DP, Mathus-Vliegen EM, Dagli U, et al. Effect of prolonged gastric distention on lower esophageal sphincter function and gastroesophageal reflux. *Am J Gastroenterol*. 2003;98:1696–1704.
71. Barrichello S, Badurdeen D, Hedjoudje A, et al. The effect of the intra-gastric balloon on gastric emptying and the DeMeester score. *Obes Surg*. 2020;30:38–45.
72. Kumar N, Bazerbachi F, Rustagi T, et al. The influence of the orbera intragastric balloon filling volumes on weight loss, tolerability, and adverse events: a systematic review and meta-analysis. *Obes Surg*. 2017;27:2272–2278.
73. Kim SY. The effect of endoscopic bariatric and metabolic therapies on gastroesophageal reflux disease. *Medicina (Kaunas)*. 2021;57.
74. Tolone S, Savarino E, de Bortoli N, et al. Esophageal high-resolution manometry can unravel the mechanisms by which different bariatric techniques produce different Reflux exposures. *J Gastrointest Surg*. 2020;24:1–7.
75. Sullivan S, Edmundowicz SA, Thompson CC. Endoscopic bariatric and metabolic therapies: new and emerging technologies. *Gastroenterology*. 2017;152:1791–1801.
76. Sharaiha R. Managing obesity with endoscopic sleeve gastroplasty. *Gastroenterol Hepatol*. 2017;13:547–549.
77. Thompson DTHdM, Eduardo Guimarães Hourneaux de M, Christopher C. Endoscopic sleeve gastroplasty: whence we came where we are going. <http://www.wjgnet.com/>; 2019.
78. Stier C, Chiappetta S. Endoluminal revision (OverStitch (TM), Apollo Endosurgery) of the dilated gastroenterostomy in patients with late dumping syndrome after proximal Roux-en-Y gastric bypass. *Obes Surg*. 2016;26:1978–1984.
79. Espinet-Coll E, Nebreda-Duran J, Galvao-Neto M, et al. Suture pattern does not influence outcomes of endoscopic sleeve gastroplasty in obese patients. *Endosc Int Open*. 2020;8:E1349–E1358.
80. Singh S, Hourneaux de Moura DT, Khan A, et al. Safety and efficacy of endoscopic sleeve gastroplasty worldwide for treatment of obesity: a systematic review and meta-analysis. *Surg Obes Relat Dis*. 2020;16:340–351.
81. Kumar N, Abu Dayyeh BK, Lopez-Nava, Breviere G, et al. Endoscopic sutured gastroplasty: procedure evolution from first-in-man cases through current technique. *Surg Endosc*. 2018;32:2159–2164.
82. Hedjoudje A, Abu Dayyeh BK, Cheskin LJ, et al. Efficacy and safety of endoscopic sleeve gastroplasty: a systematic review and meta-analysis. *Clin Gastroenterol Hepatol*. 2020;18:1043–1053. e4.
83. Jirapinyo P, Thompson CC. Endoscopic gastric body plication for the treatment of obesity: technical success and safety of a novel technique (with video). *Gastrointest Endosc*. 2020;91:1388–1394.
84. Jirapinyo P, Thompson CC. Gastric plications for weight loss: distal primary obesity surgery endoluminal through a belt-and-suspenders approach. *VideoGIE*. 2018;@ 3:296–300.
85. Singh S, Bazarbashi AN, Khan A, et al. Primary obesity surgery endoluminal (POSE) for the treatment of obesity: a systematic review and meta-analysis. *Surg Endosc*. 2022;36:252–266.
86. Fayad L, Adam A, Schweitzer M, et al. Endoscopic sleeve gastroplasty vs laparoscopic sleeve gastrectomy: a case-matched study. *Gastrointest Endosc*. 2019;89:782–788.
87. Bazerbachi F, Abu Dayyeh BK. The pressure is on! endoscopic sleeve gastroplasty vs laparoscopic sleeve gastrectomy: toward better patient allocation beyond pygmalionism. *Gastrointest Endosc*. 2019;89:789–791.

88. Braghetto I, Lanzarini E, Korn O, et al. Manometric changes of the lower esophageal sphincter after sleeve gastrectomy in obese patients. *Obes Surg*. 2010;20:357–362.
89. Shah SL, Dawod S, Dawod Q, et al. Same-session endoscopic sleeve gastroplasty and transoral incisionless fundoplication: a possible solution to a growing problem. *VideoGIE*. 2020;5:468–469.
90. Jirapinyo P, de Moura DTH, Horton LC, et al. Effect of aspiration therapy on obesity-related comorbidities: systematic review and meta-analysis. *Clin Endosc*. 2020;53:686–697.
91. Thompson CC, Abu Dayyeh BK, Kushner R, et al. Percutaneous gastrostomy device for the treatment of class II and class III obesity: results of a randomized controlled trial. *Am J Gastroenterol*. 2017;112:447–457.
92. Jirapinyo P, Kumar N, Saumoy M, et al. Association for bariatric endoscopy systematic review and meta-analysis assessing the American society for gastrointestinal endoscopy preservation and incorporation of valuable endoscopic innovations thresholds for aspiration therapy. *Gastrointest Endosc*. 2021;93:334–342. e1.

CHAPTER 8

Surgical therapy of gastroesophageal reflux disease and obesity

Matthew D. Spann and Christopher P. Menzel
Division of General Surgery, Vanderbilt University Medical Center, Nashville, TN, United States

Introduction

In recent years, we have witnessed an overwhelming rise of the obesity pandemic. In the United States, the prevalence of obesity increased nearly 50% from 2000 to 2019, while the rates of severe obesity more than doubled. Globally, the World Health Organization estimates that 39% of adults over the age of 18 were overweight, with 13% suffering from obesity. Obesity is associated with many comorbid conditions, with gastroesophageal reflux disease (GERD) being one of the most prevalent impacting 25%–40% of patients.[1] When considering treatment options for GERD, the comorbid disease of obesity must be considered as therapeutic options often differ from patients not suffering from obesity.

Proposed mechanisms for patients suffering from both diseases include dietary choices, increased mechanical stress on the diaphragmatic hiatus, and increased intragastric pressure from visceral or subcutaneous adipose tissue. The theory of increased mechanical stress is supported by the observation that patients suffering from Class I obesity (BMI 30–35) are four times more likely to have a paraesophageal hernia than patients not suffering from obesity.[2] Furthermore, large cohort studies have observed a high percentage of patients with both GERD and obesity with a defective lower esophageal sphincter (LES).[3] With respect to GERD, medical management most often with proton pump inhibitor remains the first-line therapy. Similar to patients suffering from GERD alone, patients suffering from obesity and GERD will seek antireflux surgery (ARS) for a variety of reasons including medically refractory GERD, Barrett's esophagus (BE), and nonadherence to medical therapy; however, the approach to ARS between the two groups will often diverge.

Obesity and Esophageal Disorders
DOI: https://doi.org/10.1016/B978-0-323-98365-5.00008-5

The diagnostic work-up for patients with both GERD and obesity follows traditional practice. Patients presenting for consideration of ARS should and often have already undergone upper gastrointestinal fluoroscopic imaging and esophagogastroduodenoscopy. Esophageal pH assessment will be helpful in delineating GERD from other causes of dyspepsia. Patients will benefit from esophageal motility study to help choose the appropriate sphincter augmentation. In addition, patients should be screened for gastroparesis with providers having a low threshold to obtain a gastric emptying study.

Traditional antireflux surgery

The tenants of ARS are based on an understanding of the physiologic factors that prevent reflux. The physiologic barriers to GERD include an adequate length of intraabdominal esophagus, the acute angle created at the gastroesophageal junction by the sling fibers of the stomach and phrenoesophageal ligament, and adequate length and tone of the LES. Other key physiologic factors include adequate gastric and esophageal motility. Thus traditional approaches to ARS include mobilization of the gastroesophageal junction to 3 cm below the esophageal hiatus, adequate narrowing of the esophageal hiatus with posterior cruroplasty, and augmentation of the LES. While these tenants are often proposed for patients suffering from obesity, consideration must be given to how traditional therapies perform in the setting of both GERD and obesity.

Fundoplication and paraesophageal hernia repair

Restoration of the gastroesophageal junction into the abdomen via paraesophageal hernia repair remains a mainstay of therapy in patients suffering from both GERD and obesity. The traditional next step in ARS, augmentation of the LES with a short segment fundoplication, may be ineffective and potentially worsen symptoms of dysphagia. While up to 10% of patients undergoing fundoplication will require re-operation, obesity is the biggest risk factor for failure, suffering with obesity is the biggest risk factor for failure, suffering five times more recurrence of paraesophageal hernia or failure of fundoplication. Reasons for failure include recurrent paraesophageal hernia, disrupted or malpositioned fundoplication, or both. When considering fundoplication for patients with BMI above 30, careful assessment of intraabdominal adiposity is essential. Patients suffering from central obesity pose technical challenges as the phrenoesophageal ligament extends further along the gastric cardia, making assessment

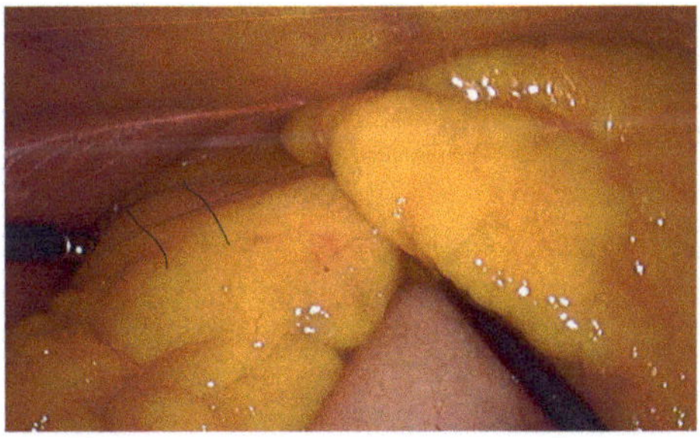

Figure 8.1 Phrenoesophageal ligament in patient suffering from morbid obesity. Green arrows represent the width of phrenoesophageal ligament. Black lines mark the edge of esophagus.

of the fundoplication difficult as shown in Fig. 8.1. While data exploring endoscopic fundoplication are limited in patients suffering from obesity, the potential for increased complications with this therapy can be extrapolated from observations after laparoscopic fundoplication. Thus patients and providers should proceed with caution when considering any fundoplication as GERD therapy for patients with BMI greater than 30. Obesity should be considered a modifiable risk factor and attempts at dietary, lifestyle, and pharmacotherapy should be considered for patients with BMI between 30 and 40. For patients with BMI greater than 40, or if attempts at nonsurgical weight loss are unsuccessful, addressing both GERD and obesity with gastric bypass should be considered and will be discussed further in this chapter.

Lower esophageal sphincter augmentation

Alternative methods of LES augmentation have been studied over many years. Currently, two devices are approved by the FDA for the treatment of GERD aimed at augmenting the LES: radiofrequency sphincter augmentation (RSA) (Stretta, Restech) and magnetic sphincter augmentation (MSA) (Linx Reflux Management, Johnson and Johnson Medical Devices). Diagnostic work-up for each procedure will be similar to other ARS. Each device has additional considerations. Decision for therapy in each device should follow careful evaluation by an experienced GERD specialist.

Radiofrequency sphincter augmentation

RSA has been approved for use by the FDA as GERD therapy since 2000. RSA is an endoscopic procedure that involves the delivery of radiofrequency treatments through a specialized balloon. During the procedure, radiofrequency energy is delivered through a series of needles that transverse the esophageal mucosa starting 2 cm above the Z line and extending 2 cm below the Z line. The exact mechanism of action is unknown but is suspected to be due to tissue remodeling within the submucosal layers. While RSA often provides a symptomatic improvement, the impact on esophageal pH is limited. Unfortunately, GERD symptom improvement after RSA in patients suffering from obesity is less, with over 40% having no improvement after treatment.[4] While RSA may not be an adequate option for ARS in patients with both GERD and obesity, it is being considered for esophageal reflux after sleeve gastrectomy and other metabolic bariatric surgeries. As esophageal reflux after weight loss surgery can be multifactorial, extensive work-up is needed before RSA. In addition, larger studies are needed to gage the long-term impact of RSA on post-bariatric esophageal reflux.

Magnetic sphincter augmentation

MSA has also been approved by the FDA as a treatment for GERD in patients that have abnormal relaxation or a hypotensive LES. MSA is a laparoscopic procedure that involves the repair of a paraesophageal hernia, if present, and then the placement of a ring comprised magnetic beads around the lower esophagus. The magnets expand with food or liquid bolus and contract after the passage of the bolus to increase LES tone. When compared to fundoplication, MSA offers benefits of decreased rates of gas/bloat symptoms as patients are typically able to both belch and vomit. Unfortunately, BMI > 35 has been associated with worse outcomes after MSA for primary reflux.[5] While exact mechanisms of decreased performance in patients suffering from obesity are unknown, increased intragastric pressure from mechanical compression is possible. MSA has recently been proposed to treat esophageal reflux after weight loss surgery. However, more studies are needed to help define patient characteristics that may benefit from this therapy.

Metabolic and bariatric surgery in patients suffering from gastroesophageal reflux disease

Bariatric and metabolic surgery offers a surgical cure for obesity and many related comorbidities, but still GERD presents a unique challenge. At baseline, multiple studies have demonstrated an increased pressure gradient across the LES as well as decreased esophageal transit times in obese individuals as opposed to normal-weight patients.[6,7] Although weight loss commonly improves symptoms of GERD, the degree to which GERD improves is unique to each patient and varies tremendously with the choice of weight-loss operation. Over the past 2 decades, we have accumulated a thorough amalgam of data by which we can better choose weight-loss operations for patients. We have written this chapter to outline our methods in tailoring the available bariatric procedures to each patient in the interest of providing optimal weight loss, resolution of obesity-related comorbidities, and maximizing treatment of existing GERD while minimizing de novo GERD from occurring.

Roux-en-Y gastric bypass

Roux-en-Y gastric bypass (RYGB) is the gold standard metabolic operation. This title reflects its enduring success in sustained weight loss and improvement of obesity-related comorbidities. Historically, RYGB was developed as a version of the Billroth II operation as an attempt to cure peptic ulcer disease without entailing the morbidity and mortality of removing the stomach. It worked well in treating symptomatic acid reflux and caused weight loss as an, initially, unintended consequence. Since that time, the RYGB has become the paramount metabolic operation and subsequently has excellent literature supporting its impact on GERD resolution. RYGB is currently the most effective bariatric procedure for the resolution of GERD and prevention of GERD-related complications, including erosive esophagitis, BE, and esophageal adenocarcinoma (EAC).[8,9]

The mechanisms by which RYGB works to resolve GERD are multiple. First by creating a gastric pouch that is small (usually no more than 5 cm in length) and excluding the fundus, majority of the acid-producing cells of the fundus and antrum are separated from the new alimentary limb. This creates a relatively low acid environment within the pouch, and after Roux-en-Y reconstruction, the acidic contents of the excluded stomach are unable to reach the esophagogastric junction. This mechanism is the same for the avoidance of alkaline reflux that can come from the proximal duodenum and cause similar symptoms and complications as GERD. Second, when we bring the Roux limb of the small intestine cephalad to connect with the pouch, the anastomosis creates a low-pressure system allowing ingested gastric contents to readily pass into the small intestine without delayed transit.[10] In addition to these anatomic changes that happen with RYGB, excess body weight loss is excellent and typically ranges from 70% to 80% as soon as 12 months after surgery. This contributes to decreased visceral and truncal obesity and less resultant pressure against the hiatus.

Many studies have demonstrated significant improvement in GERD after RYGB. Perry et al. performed RYGB on 57 obese patients who were diagnosed with GERD. All patients were reported to have sustained clinical improvement, defined as no symptoms, at a mean of 18 months follow-up. Similarly, Frezza et al. performed RYGB on 152 patients with GERD and found a statistically significant improvement in GERD symptoms ($p < 0.01$).[11]

Maldalasso et al. studied patients at longer term follow-up of initially 6 months and then at 39 months. They found statistically significant improvement of GERD symptoms as well as reflux esophagitis and esophageal acid exposure, as determined by DeMeester scores.[12] More recently, a prospective trial compared the incidence of GERD after RYGB and after sleeve gastrectomy and used DeMeester scores to quantify the difference for each group. They found the mean DeMeester score for the sleeve gastrectomy group raised from 10.9 to 40.2, which represented a statistically significant increase, whereas there was not a significant change in the RYGB group.[13]

Furthermore, one of the foremost complications of longstanding GERD, BE, has shown significant percentages of regression after RYGB. Gorodner et al. reported remission rates of 36% (4/11) of BE after RYGB. Although the numbers of preoperative BE were relatively small in this group, the average follow-up was 41 months with EGD surveillance and there were zero cases of BE that progressed to dysplasia after RYGB.[14] Andrew et al. similarly found endoscopic and histologic regression of BE in 42.9% (6 of 14) of patients after RYGB.[15] These studies emphasize the important role that RYGB plays in the treatment of BE and in preventing the progression of BE to EAC.

From curing preoperative GERD and preventing de novo postop GERD to supporting regression of BE to prevent the development of EAC, the RYGB remains the gold standard bariatric operation in the management of GERD and its related complications.

Vertical sleeve gastrectomy

Laparoscopic vertical sleeve gastrectomy (VSG) is the most common bariatric operation performed in the world. In 2021 the American Society for Metabolic & Bariatric Surgery (ASMBS) estimates that VSG accounts for 60% of primary weight-loss operations in the United States. It is commonly preferred by patients and many surgeons alike due to ease of explanation and comprehension, as well as its overall success in providing durable weight loss and metabolic improvements. As opposed to the extremely favorable data in regard to GERD resolution after RYGB, however, some studies have suggested inferior outcomes of GERD resolution as well as the increased prevalence of related complications after VSG. Although the data are not conclusive, many bariatric surgeons resultantly practice temperance when considering VSG in patients with symptomatic GERD.

The literature addresses many derivatives of GERD outcomes and related complications after VSG. The most apropos of these include the incidence of de novo GERD after VSG and resolution of pre-VSG symptomatic GERD. De novo GERD has been implicated in 8%–30% of patients following VSG.[16,17] The mechanisms by which GERD may occur de novo after VSG include reduced stomach volume and decreased gastric compliance, decreased gastric emptying, blunting of the angle of His fibers at the proximal transection point, decreased LES pressure, herniation of the proximal sleeve into the mediastinum or presence of hiatal hernia, retained fundus, and stenosis or acute angulation at the incisura angularis.[18] Many of these mechanisms can be addressed at the time of surgery while others cannot or do not develop until patients are well-removed from surgery. This fact adds to the heterogeneity of presentations and the abundance of derivations that accompany GERD.

The first *meta*-analysis by Oor et al. included 33 studies with 8,092 patients undergoing VSG.[19] Despite the large sample size they were unable to provide uniform consensus due to the variation of methods and definitions within their included studies. Still, the pooled incidence of de novo GERD was found to be 20%, which prompted the conclusion that there was an increased prevalence of symptomatic GERD post-VSG.[19] Another systematic review from 2011 showed similar ambiguity, with 4 of 15 studies publishing an increase in GERD symptoms while 7 showed a decrease in GERD symptoms following VSG.[20] Other randomized studies have shown similar de novo GERD case rates of 21.8% at 1 year status post VSG. Interestingly, in this study, only 3.1% had symptomatic GERD at three-year follow-up.[21] This longer-term resolution of GERD suggests that the weight loss after VSG, and concomitant decrease in central obesity and abdominal pressure on the hiatus, may negate some of the immediate impacts that VSG has on the development of de novo GERD.

A large retrospective study of 4832 patients found that among patients with preoperative GERD (44.5%), 84.1% of them maintained symptoms of GERD after surgery. Among the remaining 55.5% who lacked GERD symptoms before VSG, 8.6% developed de novo GERD symptoms. Studies such as this one highlight that even if the development of de novo GERD is in a minority of patients, there is minimal resolution of preexisting GERD.[22] For good reason, statistics like this give bariatric surgeons pause in recommending VSG in a patient who presents with preoperative symptomatic GERD.

There are multiple studies that support the minimization or elimination of GERD after VSG as well. In the same studies showing a 21.8% de novo incidence of GERD after VSG, Himpens et al. found that 75% of patients with preexisting GERD had resolution after VSG.[21] In addition to weight loss as a factor contributing to the resolution of GERD, other mechanisms after VSG including resection of acid-producing fundic cells and increased gastric emptying may contribute.

Another important anatomic piece of the GERD-VSG conundrum is hiatus integrity. As discussed earlier, intrinsic LES function is a major determining factor in the pathogenesis of GERD. The anatomic integrity of the hiatus, and preservation of function of the diaphragmatic crura in providing extrinsic pressure in support of LES function, is obliterated in hiatal hernias. The risk of having a hiatal hernia at the time of bariatric surgery has been observed to be as high as 25%. When hiatal hernias are repaired at the time of VSG, de novo reflux rates were observed to be as low as 0%. This compared to de novo GERD of 22.9% in the patients who had VSG without identification of a hiatal hernia.[23] Studies like these highlight the multifactorial etiologies of GERD and how they may be caused, or prevented, after VSG.

Symptomatic GERD frequently requires a revisional operation, even though the etiologies may be various. In a review of 89 patients who underwent conversion from VSG to RYGB, Landreneau et al. reported 42 of these patients had revisional surgery due to complications of their sleeves. The complications were refractory GERD in 17 patients, sleeve stenosis in 13 patients, and torsion or fistulae in 11 patients. In the refractory GERD subset of patients, 12 of the 17 followed up at 1 year and 9 of these had complete resolution of GERD.[24]

In our experience, we have seen many patients present as initial consultations to us with intractable GERD from a previously performed VSG. On our evaluation, the most common etiologies include hiatal hernia present or recurrent since the first operation, retained fundus that gives the sleeve a tapering shape, or stricture and "corkscrewing" of the sleeve, most commonly at the incisura angularis. These anatomic circumstances are generally amenable to surgical correction. We will study these patients preoperatively with endoscopy, imaging, and full set of data, including micro and macronutrients, to optimize patients for revisional surgery. We rule out the presence of GERD complications and offer patients conversion to RYGB with hiatal hernia repair as indicated. At the time of operation these cases typically have significant scar tissue between foregut structures and if a hiatal hernia was previously repaired the hiatal planes are similarly scarred to the esophagus and proximal sleeve. In cases of the retained fundus, continuing the dissection superiorly to the left crural pillar is necessary to successfully expose this portion of the stomach and resect it before creating our gastric pouch. In cases of acute angulation or stricture, we routinely use intraoperative endoscopy to ensure our gastrojejunostomy between the gastric pouch and Roux limb will be created on a healthy, normal stomach situated above the level of the stricture.

There are many mechanisms by which a VSG can cause de novo GERD and ample literature specifically addressing outcomes, which do vary tremendously. Still, the fact that the RYGB has excellent, substantiated data makes the authors favor the utilization of the RYGB in patients as the operation of choice for patients with severe preoperative GERD, GERD-related complications, and no contraindications to RYGB.

Adjustable gastric banding

Other bariatric procedures, most notably laparoscopic adjustable gastric bands (LAGB), have failed to meet the mark of long-term success in GERD treatment and prevention. A systematic review of 20 studies and 3307 patients by de Jong et al. revealed initial success after LAGB placement and reported a reduction in GERD symptoms from 33.7% to 7.7% postoperatively, as well as a significant decrease in PPI usage. However, there was a significant subset of patients who had not reported GERD preoperatively who developed de novo GERD status post-LAGB placement, 15%, and 29.4% who were observed to develop interval reflux esophagitis.[25]

More recently, Chen et al. performed high-resolution esophageal manometry and 24-hour esophageal pH studies on three cohorts of LAGB patients to further delineate the relationship between GERD symptoms and esophagitis. The groups were patients with preoperative reflux, symptomatic patients status post LAGB, and asymptomatic patients after LAGB placement. Their results showed that symptomatic LAGB patients had higher esophageal acid exposure, lower LES resting pressures, and impaired distal esophageal contractility compared to the other groups.

Himpens et al. found a correlation between the duration of LAGB and rates of developing GERD. As opposed to the date on VSG, their studies demonstrated a mere 8.8% of patients having reported de novo GERD symptoms at 1 year following LAGB placement, while this increased to 20.5% at 3 years.[21] As opposed to the conclusion with the VSG that weight loss and certain technical parts of the VSG support resolution of de novo reflux, the LAGB group develops symptomatic GERD over time. This is consistent with larger outcome studies that report a high prevalence of complications relating to LAGB that increase over time, including band slippage and erosion.

We do not offer LAGB to our patients anymore, from both a lack of efficacy compared with RYGB and VSG as well as the number of band-related complications we treat in our practice, including slippage, erosion, and esophageal motility problems with long-term banding. Although weight loss after LAGB does show an initial decrease in postoperative GERD, over time both weight and GERD are shown to have a high rate of recurrence. Similarly, among the newer endoscopic interventions for curing or managing obesity, the endoluminal sleeve has the best literature in regard to the improvement of GERD. The weight loss is modest compared to traditional laparoscopic bariatric surgery, but the decreased percentage of patients who experience postprocedural GERD is an area that should be studied to hopefully better elucidate the most relevant mechanisms by which GERD occurs after VSG and other procedures.

Endoluminal weight loss procedures

Various endoluminal techniques have been described as part of the minimally invasive repertoire that proceduralists may use in combating obesity. These endoscopic interventions include intragastric balloons, endoluminal vertical gastroplasty, endoscopic gastrointestinal bypass devices, and duodenal mucosal resurfacing. Intragastric balloons are space-occupying masses inflated within the stomach to induce satiety. These increase intragastric pressure and increase the pressure gradient across the LES. Due to these mechanisms, reflux is commonly reported and is rarely treated by this method of weight loss. Retrospective studies reported 6.8% de novo GERD after intragastric balloon placement. Additionally, it is reported that most patients with GERD are advised to remain on PPI therapy throughout intragastric balloon usage.[26]

Endoluminal vertical gastroplasty is performed in a few specialized centers and has more positive results in regards to GERD. Recent studies show very low levels of GERD after endoluminal sleeves. One such study showed 0% incidence of reflux symptoms in 64 patients at 1 year follow-up.[27] A second study compared endoluminal sleeve gastroplasty with laparoscopic sleeve gastrectomy at 6 months, and although endoluminal sleeves fell short by comparison in weight loss, they had significantly less patient-reported episodes of de novo GERD.[28]

Endoscopic bypass devices use tubes to extend from the proximal duodenum to mid jejunum, or alternatively from GE junction through the stomach into the mid jejunum. These as well as duodenal mucosal resurfacing techniques have not had ample evidence to this point to make a statement on efficacy, let alone GERD rates.[10]

Roux-en-Y gastric bypass for failed antireflux surgery

Standard ARS includes hiatal hernia repair and laparoscopic Nissen fundoplications (LNF) for primary treatment of GERD. ARS has been shown to be successful in normal-weight patients, but the efficacy of ARS wanes in the obese population. So much so is this dichotomy that many surgeons will advocate for their patients to be BMI < 30 before undergoing standard ARS. As we have observed in the first part of the 21st century, obesity has continued to surge and those patients unable to reach weight goals for surgery are significant. This problem is not relegated to those without foregut surgery, either. In patients who have had previous ARS, the weight gained in obesity puts more pressure on these patients' previous repairs and consequently leads to a relapse of GERD in many of these patients. The need for revisional surgery to cure renewed and obviate complications that arise, either from GERD itself or from failed ARS, is rising. In a surgeon's current armamentarium, RYGB has proven to be an effective operation to treat this increasingly complex and common revisional ARS predicament.

Outcomes of revisional RYGB from LNF have been well studied and evaluated over the past two decades. An initial study from Zainabadi et al. studied 7 patients converted from previous LNF to RYGB. Though one patient was converted to open procedure and there was a higher rate of early complications (3/7 patients), there were no surgical leaks, deaths, or complications arising specific to the technical complexity of the operation. Converting the LNF to RYGB also showed significant improvement based on the GERD health-related quality of life survey, with reduced scores from 16.7 preoperative to 4.4 postoperatively.[29]

Ibele et al. studied 14 patients who had previous LNP that they converted to RYGB. They concluded the feasibility and efficacy of RYGB for long-term weight loss and comorbidity successes were comparable to primary RYGB. They found similar results as did Zainabadi's group regarding safety, that complications are more frequent to encounter however long-term outcomes reflect the safety of this operation.[30]

More recently an even larger study by Coakley et al. reviewed 87 total patients with previous, failed fundoplication. 21 of the 87 patients had had two or more failed fundoplications and subsequent revisional foregut operations as well. After conversion to RYGB with a median follow-up of 35.8 months, 87.4% had a resolution of preoperative GERD symptoms. Additional findings were sustained excess body weight loss (> 80%) and perioperative morbidities that paralleled or improved upon from previous studies. These studies show that conversion of prior failed LNF to RYGB is a safe and successful option for patients with recalcitrant GERD after standard ARS.[31]

Conclusion

Obesity is often cited as our greatest threat in healthcare. Treating comorbidities in patients suffering from obesity is frequently complicated as traditional therapies do not produce the intended result. Without proper intervention and prevention, the United States and other countries in the world will find most of their population suffering from obesity in the coming years. Thus providers will see increasing incidence of GERD, paraesophageal hernia, and failed antireflux procedures. Continued investigation is needed to identify additional options for treating this complex pathology. Providers treating GERD must also advocate for the adequate diagnosis, treatment, and prevention of obesity.

References

1. Chang P, Friedenberg F. Obesity and GERD. *Gastroenterol Clin North Am.* 2014;43:161–173.
2. Wilson LJ, Ma W, Hirschowitz BI. Association of obesity with hiatal hernia and esophagitis. *Am J Gastroenterol.* 1999;94:2840–2844.
3. Ayazi S, Hagen JA, Chan LS, et al. Obesity and gastroesophageal reflux: quantifying the association between body mass index, esophageal acid exposure, and lower esophageal sphincter status in a large series of patients with reflux symptoms. *J Gastrointest Surg.* 2009;13:1440–1447.
4. White B, Jeansonne LO, Cook M, et al. Use of endoluminal antireflux therapies for obese patients with GERD. *Obes Surg.* 2009;19:783–787.
5. Warren HF, Brown LM, Mihura M, Farivar AS, Aye RW, Louie BE. Factors influencing the outcome of magnetic sphincter augmentation for chronic gastroesophageal reflux disease. *Surg Endosc.* 2018;32:405–412.
6. Mercer CD, Rue C, Hanelin L, Hill LD. Effect of obesity on esophageal transit. *Am J Surg.* 1985;149:177–181.
7. Pandolfino JE, El-Serag HB, Zhang Q, Shah N, Ghosh SK, Kahrilas PJ. Obesity: a challenge to esophagogastric junction integrity. *Gastroenterology.* 2006;130:639–649.
8. Ashrafi D, Osland E, Memon MA. Bariatric surgery and gastroesophageal reflux disease. *Ann Transl Med.* 2020;8:S11.
9. Thrift AP, Shaheen NJ, Gammon MD, et al. Obesity and risk of esophageal adenocarcinoma and Barrett's esophagus: a Mendelian randomization study. *J Natl Cancer Inst.* 2014;106.
10. Fass OZ, Mashimo H. The effect of bariatric surgery and endoscopic procedures on gastroesophageal reflux disease. *J Neurogastroenterol Motil.* 2021;27:35–45.
11. Frezza EE, Ikramuddin S, Gourash W, et al. Symptomatic improvement in gastroesophageal reflux disease (GERD) following laparoscopic Roux-en-Y gastric bypass. *Surg Endosc.* 2002;16:1027–1031.
12. Madalosso CA, Gurski RR, Callegari-Jacques SM, Navarini D, Mazzini G. Pereira MaS. The impact of gastric bypass on gastroesophageal reflux disease in morbidly obese patients. *Ann Surg.* 2016;263:110–116.
13. Raj PP, Bhattacharya S, Misra S, et al. Gastroesophageal reflux-related physiologic changes after sleeve gastrectomy and Roux-en-Y gastric bypass: a prospective comparative study. *Surg Obes Relat Dis.* 2019;15:1261–1269.
14. Gorodner V, Buxhoeveden R, Clemente G, Sánchez C, Caro L, Grigaites A. Barrett's esophagus after Roux-en-Y gastric bypass: does regression occur? *Surg Endosc.* 2017;31:1849–1854.
15. Andrew B, Alley JB, Aguilar CE, Fanelli RD. Barrett's esophagus before and after Roux-en-Y gastric bypass for severe obesity. *Surg Endosc.* 2018;32:930–936.
16. Boza C, Daroch D, Barros D, León F, Funke R, Crovari F. Long-term outcomes of laparoscopic sleeve gastrectomy as a primary bariatric procedure. *Surg Obes Relat Dis.* 2014;10:1129–1133.
17. Felinska E, Billeter A, Nickel F, et al. Do we understand the pathophysiology of GERD after sleeve gastrectomy? *Ann NY Acad Sci.* 2020;1482:26–35.
18. Patti MG, Schlottmann F. Gastroesophageal reflux after sleeve gastrectomy. *JAMA Surg.* 2018;153:1147–1148.
19. Oor JE, Roks DJ, Ünlü Ç, Hazebroek EJ. Laparoscopic sleeve gastrectomy and gastroesophageal reflux disease: a systematic review and *meta*-analysis. *Am J Surg.* 2016;211:250–267.
20. Chiu S, Birch DW, Shi X, Sharma AM, Karmali S. Effect of sleeve gastrectomy on gastroesophageal reflux disease: a systematic review. *Surg Obes Relat Dis.* 2011;7:510–515.

21. Himpens J, Dapri G, Cadière GB. A prospective randomized study between laparoscopic gastric banding and laparoscopic isolated sleeve gastrectomy: results after 1 and 3 years. *Obes Surg*. 2006;16:1450–1456.
22. DuPree CE, Blair K, Steele SR, Martin MJ. Laparoscopic sleeve gastrectomy in patients with preexisting gastroesophageal reflux disease: a national analysis. *JAMA Surg*. 2014;149:328–334.
23. Soricelli E, Iossa A, Casella G, Abbatini F, Calì B, Basso N. Sleeve gastrectomy and crural repair in obese patients with gastroesophageal reflux disease and/or hiatal hernia. *Surg Obes Relat Dis*. 2013;9:356–361.
24. Landreneau JP, Strong AT, Rodriguez JH, et al. Conversion of sleeve gastrectomy to Roux-en-Y gastric bypass. *Obes Surg*. 2018;28:3843–3850.
25. de Jong JR, Besselink MG, van Ramshorst B, Gooszen HG, Smout AJ. Effects of adjustable gastric banding on gastroesophageal reflux and esophageal motility: a systematic review. *Obes Rev*. 2010;11:297–305.
26. Al-Momen A, El-Mogy I. Intragastric balloon for obesity: a retrospective evaluation of tolerance and efficacy. *Obes Surg*. 2005;15:101–105.
27. Fogel R, De Fogel J, Bonilla Y, De La Fuente R. Clinical experience of transoral suturing for an endoluminal vertical gastroplasty: 1-year follow-up in 64 patients. *Gastrointest Endosc*. 2008;68:51–58.
28. Fayad L, Adam A, Schweitzer M, et al. Endoscopic sleeve gastroplasty vs laparoscopic sleeve gastrectomy: a case-matched study. *Gastrointest Endosc*. 2019;89:782–788.
29. Zainabadi K, Courcoulas AP, Awais O, Raftopoulos I. Laparoscopic revision of Nissen fundoplication to Roux-en-Y gastric bypass in morbidly obese patients. *Surg Endosc*. 2008;22:2737–2740.
30. Ibele A, Garren M, Gould J. The impact of previous fundoplication on laparoscopic gastric bypass outcomes: a case-control evaluation. *Surg Endosc*. 2012;26:177–181.
31. Coakley KM, Groene SA, Colavita PD, et al. Roux-En-Y gastric bypass following failed fundoplication. *Surg Endosc*. 2018;32:3517–3524.

CHAPTER 9

Postbariatric surgery esophageal dysmotility

Joshua Lee[1], Benjamin Lloyd[1], Joseph Wawrzynski[2] and Amit Patel[3]
[1]Division of Gastroenterology, Duke University School of Medicine, Durham, NC, United States
[2]Department of Medicine, Duke University School of Medicine, Durham, NC, United States
[3]Division of Gastroenterology, Duke University School of Medicine and the Durham Veterans Affairs Medical Center, Durham, NC, United States

Introduction

Obesity, defined as a body mass index (BMI) $\geq 30\,kg/m^2$, represents a growing epidemic with a worrisome estimated prevalence in the United States of 35%–41%.[1] Obesity is frequently accompanied by multiple comorbidities that can include hypertension, diabetes, cardiovascular disease, obstructive sleep apnea, and fatty liver disease.[2] From an esophageal perspective, obesity appears to correlate with increased symptom burden that may stem from increased levels of esophageal acid exposure, but not necessarily from the presence of esophageal motor disorders.[3] The efficacy of lifestyle-based and pharmacologic approaches for durable weight loss may be limited for large proportions of obese patients, particularly in the longer term, and bariatric surgical options may be considered.[4] There have been significant increases in the utilization of bariatric procedures, with over 6,00,000 such interventions performed on an annual basis worldwide.[5,6] With this increased utilization of bariatric surgery, there is increasing recognition of how these surgical interventions can affect esophageal motor function and the development of dysphagia. In fact, dysphagia represents a commonly encountered long-term complication of bariatric surgery, with the presence of postoperative esophageal dysmotility.[7]

In this chapter, we briefly review commonly utilized bariatric surgical interventions. We then describe potential approaches to the evaluation of dysphagia in postbariatric surgery patients, and the specific effects that these surgeries can have on esophageal motor function. We review data on management options and special considerations in these patients, as well as potential opportunities for better understanding.

Obesity and Esophageal Disorders
DOI: https://doi.org/10.1016/B978-0-323-98365-5.00003-6

Bariatric surgical procedures

Commonly utilized bariatric interventions discussed here include laparoscopic adjustable gastric banding (LAGB), laparoscopic sleeve gastrectomy (LSG), Roux-en-Y gastric bypass (RYGB), biliopancreatic diversion with duodenal switch (BPD-DS), and intragastric balloons (IGB), which afford effective weight loss options for obese patients (Fig. 9.1).[8] The American Society for Metabolic and Bariatric Surgery Numbers Taskforce reported that 2,52,000 such procedures were performed in 2018 in the United States alone, representing a 10.8% increase from 2017, and a 60% increase since 2011.[9] Among these bariatric procedures performed in 2018, LSG accounted for 61.4%, followed by RYGB (17%), IGB (2.0%), LAGB (1.1%), and BPD-DS (0.8%). In contrast, the most commonly utilized procedures in 2011 were RYGB (36.7%) and LAGB (35.4%).[9]

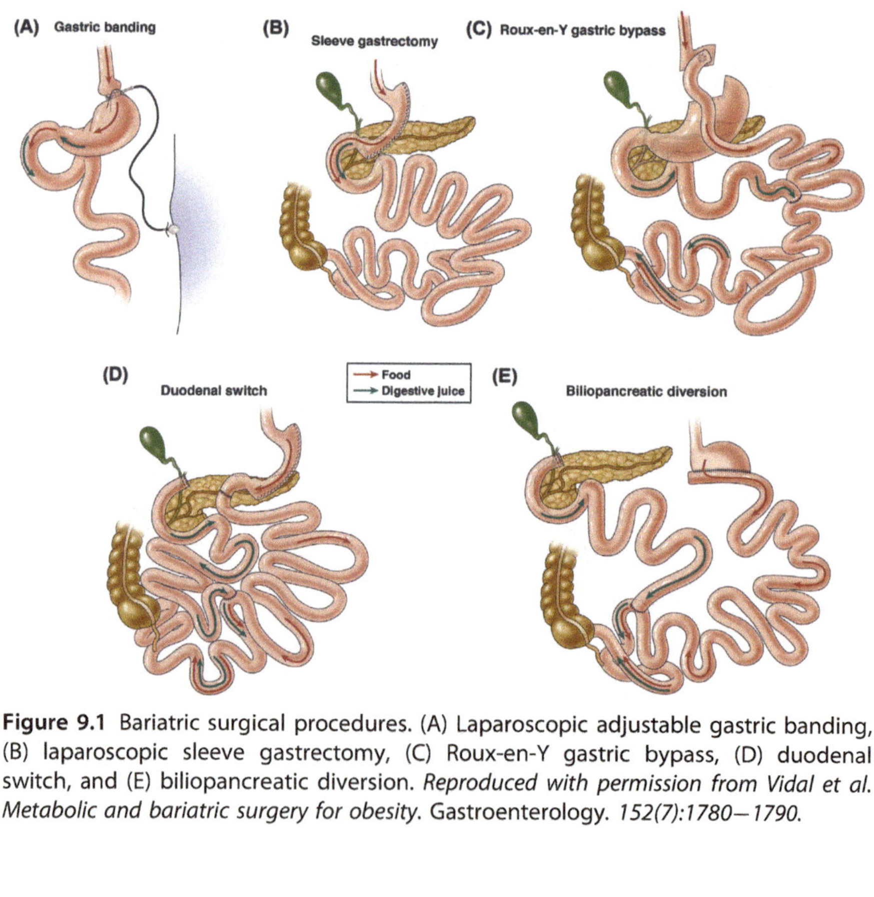

Figure 9.1 Bariatric surgical procedures. (A) Laparoscopic adjustable gastric banding, (B) laparoscopic sleeve gastrectomy, (C) Roux-en-Y gastric bypass, (D) duodenal switch, and (E) biliopancreatic diversion. *Reproduced with permission from Vidal et al. Metabolic and bariatric surgery for obesity.* Gastroenterology. *152(7):1780–1790.*

LAGB, first described in the early 1990s, involves surgical placement of an adjustable silicone band around the proximal stomach (Fig. 9.1A). This intervention creates a smaller gastric pouch proximal to the band, thereby leading to weight loss by restricting food capacity.[10] The band is connected to a subcutaneous port that allows for inflation of the band and further adjustment in the size of the gastric reservoir. While this technique was initially commonly utilized in the setting of shorter operative and recovery times, decreased relative morbidity, and the potential for reversibility, LAGB has since been utilized less frequently among bariatric providers. This change has occurred due to concerns about the efficacy and durability of weight loss, complications of band erosion and/or migration, and the presence of increasing alternative bariatric interventional options.[11]

LSG has now replaced LAGB as the most widely performed bariatric surgery due to technical ease with the evolution of laparoscopic techniques, as well as effective, durable weight loss. LSG is performed by resecting the majority (about three-fourths) of the stomach, with a vertical gastrectomy from the gastroesophageal junction to the distal antrum, thereby creating a tubular gastric pouch with preservation of the pylorus (Fig. 9.1B). Like LAGB, LSG is a primarily restrictive intervention, limiting the amount of food that the stomach can hold.

In contrast, RYGB leads to weight loss through both restrictive and malabsorptive means. RYGB is characterized by the creation of a small proximal gastric pouch, with an anastomosis to a Roux limb (loop of jejunum) to limit absorptive capacity due to delayed exposure of food to pancreatic enzymes and bile (Fig. 9.1C). Evolving from prior experiences with partial gastrectomy and Billroth II surgeries, the Roux limb is typically 75−150 cm in length, with an enteroenterostomy connecting it to the biliopancreatic limb (which is in continuity with the excluded portion of the stomach).[12]

While BPD-DS interventions afford great potential for weight loss, they are among the less frequently performed bariatric interventions at present due to longer operative times, higher risks of complications, and increased postoperative nutritional derangements.[13] BPD-DS surgeries are characterized by the removal of part of the stomach as well as bypassing of a significant portion of the small intestine (Fig. 9.1D,E). Like RYGB, BPD-DS interventions combine both restrictive and malabsorptive mechanisms to achieve weight loss.

IGB options have arrived more recently, with the benefits of representing nonsurgical procedures that are widely available, reversible, and associated with lower rates of adverse events.[14] Specifically, currently available IGB options may be single or multiple, filled with saline or air, and can be endoscopically placed or orally ingested, based on the device. The balloons are restrictive in nature by occupying space within the stomach, thereby leading to altered satiety and delayed gastric emptying.[14] IGB devices are designed to be temporary in nature, with dwell times that may be around 6 months before balloon removal. Overall weight loss may be more modest with IGB than the surgical alternatives mentioned here, though positive metabolic outcomes have been reported.[15]

Evaluation of dysphagia after bariatric surgery

Dysphagia represents a common long-term complication after bariatric surgical approaches. Multicenter American data reported as a substudy of the Longitudinal Assessment of Bariatric Surgery Consortium focused on surgery-related GI symptoms at 1–3 years after LAGB and RYGB, revealing high rates of dysphagia among these patients.[16] Specifically, based on self-reported symptoms, the prevalence of dysphagia symptoms at least once weekly was 43.9% and 16.7% at 1 year and 27.5% and 10.9% at 3 years after LAGB and RYGB, respectively. Among a separate cohort of 271 patients from three large tertiary referral sites (mixed population of LSG and RYGB), the prevalence of dysphagia was 13.7% at a mean follow-up of 3.9 years after bariatric surgery.[7]

Given these high rates of dysphagia after bariatric surgery, a systematic approach is appropriate for the evaluation of symptoms. As with most other patients who present with dysphagia, upper endoscopy typically represents an initial investigation to evaluate for mucosal and/or structural obstructive processes, such as strictures, esophagitis, or masses, as well as with biopsies to rule out eosinophilic esophagitis.[17] Further esophageal physiologic tests (such as barium esophagography, high-resolution manometry (HRM), endoscopic functional lumen imaging probe (FLIP), mucosal integrity) may be utilized, though a careful symptom history is indicated in these settings to guide evaluation.[18] Specifically, the type of bariatric intervention performed and the timing of dysphagia symptoms can help guide a thoughtful and directed work-up.

In the case of LAGB, the acute onset of obstructive symptoms, which may include nausea, vomiting, and pain with oral intake soon after surgery, may result from narrowing of the gastric stoma due to slippage (anterior or posterior), gastric prolapse, or postsurgical edema.[19,20] Over the longer term, banding can also lead to esophageal dilation and a pseudoachalasia phenomenon, that may be reversible, as discussed later in this chapter.[21,22] Imaging with barium studies may be helpful to confirm band position and contrast transit. If the band is appropriately positioned but suspected to contribute to symptoms, band deflation may be considered, before band removal or replacement.[20]

In the setting of RYGB, upper endoscopy and/or barium imaging may be particularly useful for the identification of anastomotic strictures that can present with dysphagia in the early postoperative setting. Among a cohort of 101 patients from the Scandinavian Obesity Surgery Registry with anastomotic strictures within the first postoperative year after RYGB, 93% of stricture patients complained of dysphagia, with a median time from RYGB to stricture diagnosis of 51 days; 75% of these strictures were diagnosed within 70 days of surgery.[23] In this cohort, risk factors for strictures included patient age >60 years, circular stapled gastrojejunostomy, postoperative anastomotic leak, and marginal ulceration.[23] Chilean data reported 23% and 36% of patients had anastomotic strictures found on routine endoscopy performed 1 month after open (with circular stapler for anastomosis) or laparoscopic (with one hand-sewn continuous suture for anastomosis) RYGB, respectively.[24]

If endoscopy is unrevealing for a mucosal or structural process generating dysphagia symptoms, esophageal HRM should be considered for evaluation of esophageal motor function. HRM is particularly appropriate in this postbariatric surgery setting given the potential effects of bariatric surgery on esophageal motility, as discussed in the following section.[18,25,26] On the other hand, in the absence of structural abnormalities such as strictures on endoscopy, the utility of barium esophagography is limited, with suboptimal sensitivity and specificity for detecting esophageal dysmotility (based on HRM as the gold standard) of 69% and 50%, respectively.[27] The Chicago Classification, currently in its fourth iteration, proposes the performance, nomenclature, and interpretation of esophageal HRM studies, incorporating provocative maneuvers into the standardized HRM protocol.[28] These maneuvers should include upright wet swallows, to better arbitrate potential esophagogastric junction outflow obstruction (EGJOO), and the rapid drink challenge, to detect latent EGJ outflow obstruction.[29]

The endoscopic functional lumen imaging probe (FLIP) is a novel technological application that evaluates the mechanical properties of the esophageal lumen and EGJ.[30,31] FLIP is performed with a balloon-mounted catheter utilized at sedated upper endoscopy to acquire data on distensibility indices and contractile activity.[32] FLIP can be useful in the work-up of dysphagia, particularly in settings of clinically suspected achalasia spectrum disorders but equivocal or borderline HRM findings, and potentially during or after intervention for achalasia.[18,31,33] There are limited data on the use of FLIP specifically in the postbariatric surgery setting, but pilot data do support the utility of FLIP in the characterization and management of gastric sleeve stenosis, with symptom responders demonstrating larger postdilation diameters and distensibility indices compared to nonresponders.[34] Further work has sought to utilize FLIP to help develop quantifiable endoscopic criteria to characterize gastric sleeve stenosis.[35]

Effects of bariatric surgery on esophageal motor function

Multiple lines of investigations have revealed that bariatric interventions, whether LAGB, LSG, RYGB, BPD-DS, or IGB, may be associated with changes in esophageal motor dynamics (Table 9.1).[36] LAGB is associated with higher risks of impaired lower esophageal sphincter (LES) relaxation and ineffective esophageal motility (IEM) and has the highest frequency of postoperative esophageal motor disorders among bariatric interventions.[37–39] An older comparative study of 43 patients with esophageal testing before LAGB and 6–18 months after LAGB revealed weaker lower esophageal contractions (significant decreases in lower esophageal contraction amplitudes from 94.3 to 66.0 mmHg) and a trend toward more postoperative nonspecific motility disorders.[40] A systematic review investigating the effects of LAGB on esophageal motility reported increases in LES pressures from 12.9 to 16.9 mmHg and decreases in LES relaxation from 100% to 79.7%, with increases in dysmotility from 3.5% to 12.6%.[41] In a prospective trial of 167 patients who underwent LAGB followed with annual barium swallows for up to 12 years, 68.8% had evidence of esophageal dysmotility and 25.5% developed esophageal dilatation (defined as an esophageal diameter ≥35 mm).[42]

In terms of comparison with RYGB, a Swiss study found higher proportions of patients with esophageal barium contrast retention of >1 cm at 30 seconds after LAGB (73%) than after either direct RYGB (naïve to bypass, 9%) or after conversion from LAGB to RYGB (21%). In this cohort, LAGB patients not only had higher rates of dysphagia, chest pain, and regurgitation symptoms, but also higher proportions of ineffective contractions on liquid swallows as well as incomplete bolus transit (based on impedance) on liquid and viscous swallows.[43] An Italian study

Table 9.1 Summary of reported potential effects of bariatric procedures on esophageal motility.

Bariatric surgical procedure	Reported potential effects on esophageal motility
Gastric banding	• *Increased* LES pressures, EGJ contractile integral, and incomplete LES relaxation[44,60] • *Decreased* esophageal contraction amplitudes, with *increased* proportions of ineffective contractions and incomplete bolus transit[40,43,60,61] • *Increased* IGP and GEPG[44] • *Possible* esophageal dilation and pseudoachalasia patterns[50,51,53]
Sleeve gastrectomy	• *Decreased* LES resting pressures[45–47] • *Decreased* esophageal contractile vigor and *increased* frequency of ineffective peristalsis[44,45,47] • *Increased* IGP and GEPG[44]
Roux-en-Y gastric bypass	• *Possibly increased* rates of ineffective esophageal motility, but without consistent changes in overall esophageal body amplitudes[47,62,63] • *Unchanged* LES pressures or length[47,62] • *Decreased* IGP and GEPG[44] • *Possible* achalasia or postobesity surgery esophageal dysfunction (POSED) patterns[7]
Biliopancreatic diversion-duodenal switch	• *Unchanged* LES pressures[44] • *Decreased* IGP and GEPG[44]
Intragastric balloons	• *Unchanged* LES pressures, IGP, and GEPG[44,49]

EGJ, esophagogastric junction; *GEPG*, gastroesophageal pressure gradient; *IGP*, intragastric pressure; *LES*, lower esophageal sphincter.

investigated a mixed cohort of over 100 obese patients (who underwent different bariatric interventions, including LAGB, LSG, RYGB, BPD, IGB) with esophageal diagnostics (including HRM) performed preoperatively and 1 year postoperatively.[44] Regarding changes in EGJ metrics (LES pressure, integrated relaxation pressure (IRP), and EGJ-contractile integral), only LAGB among the bariatric interventions demonstrated significant changes (with significant increases across all three of these EGJ metrics postoperatively), though smaller sample sizes and limited follow-up durations limited this analysis.

While LSG has been known to affect gastric motility, data suggest that esophageal motor function may also be affected. An older systematic review noted increases in IEM diagnosis and reductions in contractile vigor after LSG, along with worsening of bolus transit compared to baseline.[45] Further, 7 of 12 studies in this older systematic review noted decreases in LES tone. Interestingly, among a cohort of Chilean patients who underwent LSG and developed esophageal symptoms postoperatively, mean LES resting pressures decreased significantly from preoperative testing to testing performed 6 months after LSG (15.2–8.3 mmHg), a phenomenon not observed for those few patients without esophageal symptoms in this cohort.[46] However, a newer systematic review and meta-analysis of esophageal motor changes after bariatric surgery, including over 600 patients with LSG, found significantly lower LES pressures (but not length), decreased esophageal body amplitudes, and increased risks of ineffective peristalsis, after LSG.[47]

Prior thinking presumed that RYGB had limited effects on esophageal motor function, rendering it a reasonable option for those obese patients with known esophageal dysmotility.[38] Recent meta-analysis demonstrated an increased rate of IEM after RYGB, but without significant changes in LES pressures, LES length, or esophageal body amplitudes.[47] A newer study from the three Mayo Clinic centers investigated nearly 100 patients who underwent diagnostic HRM at a median of 5.8 years after LSG or RYGB, as well as a preoperative bariatric surgery cohort.[7] Among those patients who had undergone RYGB, this study found that 15.5% had either achalasia or postobesity surgery esophageal dysfunction (POSED), 8.6% with achalasia, and 6.9% with POSED, compared to none of the patients in the preoperative comparison cohort. These proportions were numerically (but not statistically significantly) higher than for those patients who had undergone LSG. POSED is best characterized as an achalasia-like pattern with complete esophageal aperistalsis and increased intragastric pressures (IGP) >30 mmHg, but with a normal IRP. The authors of this study speculated that POSED may potentially develop from a high-pressure gastric zone in the noncompliant post-RYGB stomach (and for LSG, from increased esophageal afterload acting in an obstructive fashion). On the other hand, vagus nerve injury at bariatric surgery may represent a possible pathophysiologic mechanism contributing to some cases of achalasia.[48] Regardless, these Mayo Clinic study findings, especially the associations between increasing duration since surgery and the development of achalasia, POSED, or major manometric disorders highlight how esophageal motility disorders may represent underrecognized long-term complications of obesity surgery with time-dependent risk.[7]

Given their relative novelty and their limited dwell times, IGB interventions have limited data on postprocedural changes in esophageal motility. One single-center retrospective analysis of 24 patients with HRM performed before IGB placement and again with IGB in place found similar LES pressures with IGB placement (14.7 mmHg vs 17.8 mmHg); further, changes in LES pressures were not associated with changes in DeMeester scores.[49] The aforementioned Italian study with a mixed bariatric intervention cohort revealed significant increases in IGP and gastroesophageal pressure gradients (GEPG) after LAGB and LSG, but decreases after RYGB and BPD; these metrics remained similar for the subset of patients who underwent IGB.[44]

Management of dysphagia after bariatric surgery

As for patients without a prior history of bariatric surgery, if dysphagia stems from strictures, dilation represents a reasonable intervention when technically feasible. However, dilation may carry increased risks and warrant additional sessions in the setting of anastomotic strictures. Among the Scandinavian Obesity Surgery Registry cohort of patients with anastomotic strictures after RYGB, one or two dilation sessions were sufficient to successfully treat half of patients, though the risk of perforation at each dilation in this cohort was reported as 3.8%.[23] Among the Chilean data on routine endoscopy performed 1 month after RYGB, patients with mild (diameter 7−9 mm) or moderate (5−6 mm) strictures typically needed only one or two dilations, but those with critical strictures (≤4 mm) typically required three to five dilations.[24] Other mucosal or structural etiologies for dysphagia, such as reflux esophagitis, eosinophilic esophagitis, infectious esophagitis, or malignancy, should be managed as appropriate, if diagnosed at endoscopy.

From an esophageal motor perspective, as discussed earlier, esophageal dysmotility represents an underrecognized phenomenon after bariatric surgical intervention. Specifically, potential pseudoachalasia and/or esophageal dilation after LAGB may be reversible. This can typically be managed with band deflation and/or band removal, often with good clinical success and resolution of esophageal dilation on follow-up imaging.[21,22,50,51] In this setting, revision to RYGB may also be considered for weight loss, with normalization of barium swallow abnormalities and resolution of dysphagia symptoms reported at postoperative follow-up in one series.[52] Interestingly, one study found that esophageal dilatation and dysmotility on a routine annual barium radiography protocol were distributed similarly after LAGB regardless of the presence of symptoms. In this study, 51.3% of symptomatic patients and 47.9% of asymptomatic patients had these changes noted on an imaging study at an average follow-up of 6.8 years after LAGB.[53] Further, barium findings normalized in over half of those patients who underwent band removal and had repeat imaging performed.

Achalasia or POSED, as noted in the Mayo Clinic study, can be diagnosed after RYGB or LSG.[7] While data or consensus around the management of POSED is limited, there are multiple and growing reports of myotomy approaches utilized successfully for achalasia in the postbariatric surgery population. Given the youth of many of these patients, and the relative efficacy and durability of these achalasia interventions, intrasphincteric botulinum toxin injection or pneumatic dilation may be less ideal than surgical or endoscopic myotomy for most patients.[54] Performance of surgical Heller myotomy for patients with achalasia with a prior history of RYGB has been reported with good symptomatic outcomes; the remnant stomach may also be used for a fundoplication.[55]

Given concerns related to prior abdominal surgical intervention with potential adhesions, as well as RYGB anatomy serving to protect against postoperative GERD, there has been increasing interest in per-oral endoscopic myotomy (POEM) for the management of achalasia in this population.[56] A series of 10 achalasia patients with prior RYGB surgery underwent POEM, all with clinical success (defined as a postoperative Eckardt score ≤ 3), and a significant decrease in mean Eckardt score from 6.5 to 1, at a median follow-up duration of 1.6 years after POEM.[57] Of note, all six of the patients in this series who underwent postprocedural pH testing had normal DeMeester scores. A recently published narrative review including 12 studies and a total of 28 patients with achalasia diagnosed after RYGB found that both Heller myotomy and POEM represented feasible, safe, and effective interventions in these patients.[58] Another study described a small series of achalasia patients who had previously undergone either LSG or RYGB, for whom POEM was technically feasible and offered an effective therapeutic option.[59]

Conclusions

Given marked increases in the prevalence of obesity and the utilization of bariatric surgical interventions, careful attention should be paid to dysphagia symptoms and esophageal dysmotility, which are surprisingly common in these populations. Despite limited data to guide clinical care, special considerations are present in these patients. In particular, LAGB is associated with the greatest frequency of esophageal motor dysfunction. This can manifest as pseudoachalasia and esophageal dilatation, which may be reversible with band deflation and/or removal. While RYGB and LSG appear to have comparatively lower frequencies of esophageal motor dysfunction, achalasia or a unique entity described as POSED may develop after bariatric surgery with time-dependent risk patterns. For achalasia in these settings, Heller myotomy or increasingly POEM represents effective management options. Further investigations and inquiry are warranted to enhance our understanding of these phenomena, including the utility of preoperative esophageal motor evaluation, to improve the evaluation and management of esophageal dysmotility patterns and dysphagia after bariatric surgical interventions.

Conflicts of interest

Amit Patel serves as an Editorial Contributor for Practice Update and on the International Editorial Board of *Gastroenterology*. Joshua Lee, Benjamin Lloyd and Joseph Wawrzynski, disclose no conflicts.

References

1. Flegal KM, Kruszon-Moran D, Carroll MD, et al. Trends in obesity among adults in the United States, 2005 to 2014. *JAMA*. 2016;315:2284–2291.
2. Ogden CL, Yanovski SZ, Carroll MD, et al. The epidemiology of obesity. *Gastroenterology*. 2007;132:2087–2102.
3. Rogers BD, Patel A, Wang D, et al. Higher esophageal symptom burden in obese subjects results from increased esophageal acid exposure and not from dysmotility. *Clin Gastroenterol Hepatol*. 2020;18:1719–1726.
4. Garvey WT, Mechanick JI, Brett EM, et al. American association of clinical endocrinologists and American college of endocrinology comprehensive clinical practice guidelines for medical care of patients with obesity. *Endocr Pract*. 2016;22(Suppl 3):1–203.
5. Angrisani L, Santonicola A, Iovino P, et al. Bariatric surgery and endoluminal procedures: IFSO worldwide survey 2014. *Obes Surg*. 2017;27:2279–2289.
6. Angrisani L, Santonicola A, Iovino P, et al. IFSO worldwide survey 2016: primary, endoluminal, and revisional procedures. *Obes Surg*. 2018;28:3783–3794.
7. Miller AT, Matar R, Abu Dayyeh BK, et al. Postobesity surgery esophageal dysfunction: a combined cross-sectional prevalence study and retrospective analysis. *Am J Gastroenterol*. 2020;115:1669–1680.
8. Vidal J, Corcelles R, Jiménez A, et al. Metabolic and bariatric surgery for obesity. *Gastroenterology*. 2017;152:1780–1790.

9. English WJ, DeMaria EJ, Hutter MM, et al. American society for metabolic and bariatric surgery 2018 estimate of metabolic and bariatric procedures performed in the United States. *Surg Obes Relat Dis*. 2020;16:457–463.
10. Belachew M, Legrand M, Vincenti VV, et al. Laparoscopic placement of adjustable silicone gastric band in the treatment of morbid obesity: how to do it. *Obes Surg*. 1995;5:66–70.
11. Toolabi K, Golzarand M, Farid R. Laparoscopic adjustable gastric banding: efficacy and consequences over a 13-year period. *Am J Surg*. 2016;212:62–68.
12. Baker MT. The history and evolution of bariatric surgical procedures. *Surg Clin North Am*. 2011;91:1181–1201, viii.
13. Arterburn DE, Courcoulas AP. Bariatric surgery for obesity and metabolic conditions in adults. *Bmj*. 2014;349:g3961.
14. Bazerbachi F, Vargas EJ, Abu Dayyeh BK. Endoscopic bariatric therapy: a guide to the intragastric balloon. *Am J Gastroenterol*. 2019;114:1421–1431.
15. Bazerbachi F, Vargas EJ, Rizk M, et al. Intragastric balloon placement induces significant metabolic and histologic improvement in patients with nonalcoholic steatohepatitis. *Clin Gastroenterol Hepatol*. 2021;19(146-154).e4.
16. Kalarchian MA, King WC, Devlin MJ, et al. Surgery-related gastrointestinal symptoms in a prospective study of bariatric surgery patients: 3-year follow-up. *Surg Obes Relat Dis*. 2017;13:1562–1571.
17. Posner S, Boyd A, Patel A. Dysphagia in a 34-year-old woman. *JAMA*. 2020;.
18. Gyawali CP, Carlson DA, Chen JW, et al. ACG clinical guidelines: clinical use of esophageal physiologic testing. *Am J Gastroenterol*. 2020;115:1412–1428.
19. DeMaria EJ. Laparoscopic adjustable silicone gastric banding: complications. *J Laparoendosc Adv Surg Tech A*. 2003;13:271–277.
20. Blachar A, Blank A, Gavert N, et al. Laparoscopic adjustable gastric banding surgery for morbid obesity: imaging of normal anatomic features and postoperative gastrointestinal complications. *AJR Am J Roentgenol*. 2007;188:472–479.
21. Khan A, Ren-Fielding C, Traube M. Potentially reversible pseudoachalasia after laparoscopic adjustable gastric banding. *J Clin Gastroenterol*. 2011;45:775–779.
22. Arias IE, Radulescu M, Stiegeler R, et al. Diagnosis and treatment of megaesophagus after adjustable gastric banding for morbid obesity. *Surg Obes Relat Dis*. 2009;5:156–159.
23. Almby K, Edholm D. Anastomotic strictures after Roux-en-Y gastric bypass: a cohort study from the Scandinavian obesity surgery registry. *Obes Surg*. 2019;29:172–177.
24. Csendes A, Burgos AM, Burdiles P. Incidence of anastomotic strictures after gastric bypass: a prospective consecutive routine endoscopic study 1 month and 17 months after surgery in 441 patients with morbid obesity. *Obes Surg*. 2009;19:269–273.
25. Patel A, Posner S, Gyawali CP. Esophageal high-resolution manometry in gastroesophageal reflux disease. *JAMA*. 2018;320:1279–1280.
26. Gyawali CP, Patel A. Esophageal motor function: technical aspects of manometry. *Gastrointest Endosc Clin N Am*. 2014;24:527–543.
27. O'Rourke AK, Lazar A, Murphy B, et al. Utility of esophagram vs high-resolution manometry in the detection of esophageal dysmotility. *Otolaryngol Head Neck Surg*. 2016;154:888–891.
28. Yadlapati R, Kahrilas PJ, Fox MR, et al. Esophageal motility disorders on high-resolution manometry: Chicago classification version 4.0©. *Neurogastroenterol Motil*. 2021;33:e14058.
29. Horton A, Jawitz N, Patel A. The clinical utility of provocative maneuvers at esophageal high-resolution manometry (HRM). *J Clin Gastroenterol*. 2021;55:95–102.
30. McMahon BP, Frokjaer JB, Drewes AM, et al. A new measurement of oesophago-gastric junction competence. *Neurogastroenterol Motil*. 2004;16:543–546.
31. Dorsey YC, Posner S, Patel A. Esophageal functional lumen imaging probe (FLIP): how can FLIP enhance your clinical practice? *Dig Dis Sci*. 2020;65:2473–2482.

32. Hirano I, Pandolfino JE, Boeckxstaens GE. Functional lumen imaging probe for the management of esophageal disorders: expert review from the clinical practice updates committee of the AGA institute. *Clin Gastroenterol Hepatol.* 2017;15:325–334.
33. Carlson DA, Lin Z, Kahrilas PJ, et al. The functional lumen imaging probe detects esophageal contractility not observed with manometry in patients with achalasia. *Gastroenterology.* 2015;149:1742–1751.
34. Yu JX, Baker JR, Watts L, et al. Functional lumen imaging probe is useful for the quantification of gastric sleeve stenosis and prediction of response to endoscopic dilation: a pilot study. *Obes Surg.* 2020;30:786–789.
35. Yu JX, Dolan RD, Bhalla S, et al. Quantification of gastric sleeve stenosis using endoscopic parameters and image analysis. *Gastrointest Endosc.* 2020;93:1344–1348.
36. Tolone S, Savarino E, Yates RB. The impact of bariatric surgery on esophageal function. *Ann NY Acad Sci.* 2016;1381:98–103.
37. Savarino E, Marabotto E, Savarino V. Effects of bariatric surgery on the esophagus. *Curr Opin Gastroenterol.* 2018;34:243–248.
38. Frazzoni M, Frazzoni L, Tolone S, et al. Lack of improvement of impaired chemical clearance characterizes PPI-refractory reflux-related heartburn. *Am J Gastroenterol.* 2018;113:670–676.
39. Naik RD, Choksi YA, Vaezi MF. Consequences of bariatric surgery on oesophageal function in health and disease. *Nat Rev Gastroenterol Hepatol.* 2016;13:111–119.
40. Suter M, Dorta G, Giusti V, et al. Gastric banding interferes with esophageal motility and gastroesophageal reflux. *Arch Surg.* 2005;140:639–643.
41. de Jong JR, Besselink MG, van Ramshorst B, et al. Effects of adjustable gastric banding on gastroesophageal reflux and esophageal motility: a systematic review. *Obes Rev.* 2010;11:297–305.
42. Naef M, Mouton WG, Naef U, et al. Esophageal dysmotility disorders after laparoscopic gastric banding—an underestimated complication. *Ann Surg.* 2011;253:285–290.
43. Borovicka J, Krieger-Grübel C, van der Weg B, et al. Effect of morbid obesity, gastric banding and gastric bypass on esophageal symptoms, mucosa and function. *Surg Endosc.* 2017;31:552–560.
44. Tolone S, Savarino E, de Bortoli N, et al. Esophageal high-resolution manometry can unravel the mechanisms by which different bariatric techniques produce different reflux exposures. *J Gastrointest Surg.* 2020;24:1–7.
45. Balla A, Meoli F, Palmieri L, et al. Manometric and pH-monitoring changes after laparoscopic sleeve gastrectomy: a systematic review. *Langenbecks Arch Surg.* 2021;.
46. Braghetto I, Korn O. Late esophagogastric anatomic and functional changes after sleeve gastrectomy and its clinical consequences with regards to gastroesophageal reflux disease. *Dis Esophagus.* 2019;32.
47. Jaruvongvanich V, Matar R, Ravi K, et al. Esophageal pathophysiologic changes and adenocarcinoma after bariatric surgery: a systematic review and *meta*-analysis. *Clin Transl Gastroenterol.* 2020;11:e00225.
48. Shah RN, Izanec JL, Friedel DM, et al. Achalasia presenting after operative and nonoperative trauma. *Dig Dis Sci.* 2004;49:1818–1821.
49. Barrichello S, Badurdeen D, Hedjoudje A, et al. The effect of the intra-gastric balloon on gastric emptying and the DeMeester score. *Obes Surg.* 2020;30:38–45.
50. Roman S, Kahrilas PJ. Pseudoachalasia and laparoscopic gastric banding. *J Clin Gastroenterol.* 2011;45:745–747.
51. Robert M, Golse N, Espalieu P, et al. Achalasia-like disorder after laparoscopic adjustable gastric banding: a reversible side effect? *Obes Surg.* 2012;22:704–711.
52. Rogers AM. Improvement of esophageal dysmotility after conversion from gastric banding to gastric bypass. *Surg Obes Relat Dis.* 2010;6:681–683.

53. Friedman DT, Duffy AJ. Outcomes of routine upper gastrointestinal series screening and surveillance after laparoscopic adjustable gastric banding. *Surg Endosc.* 2020;34:2178–2183.
54. Casas MA, Schlottmann F, Herbella FAM, et al. Esophageal achalasia after Roux-en-Y gastric bypass for morbid obesity. *Updates Surg*. 2019;71:631–635.
55. Boules M, Corcelles R, Zelisko A, et al. Achalasia after bariatric surgery. *J Laparoendosc Adv Surg Tech A*. 2016;26:428–432.
56. Bashir U, El Abiad R, Gerke H, et al. Peroral endoscopic myotomy is feasible and safe in a gastric bypass population. *Obes Surg*. 2019;29:3523–3526.
57. Sanaei O, Draganov P, Kunda R, et al. Peroral endoscopic myotomy for the treatment of achalasia patients with Roux-en-Y gastric bypass anatomy. *Endoscopy*. 2019;51:342–345.
58. Aiolfi A, Tornese S, Bonitta G, et al. Management of esophageal achalasia after Roux-en-Y gastric bypass: narrative review of the literature. *Obes Surg*. 2019;29:1632–1637.
59. Kolb JM, Jonas D, Funari MP, et al. Efficacy and safety of peroral endoscopic myotomy after prior sleeve gastrectomy and gastric bypass surgery. *World J Gastrointest Endosc*. 2020;12:532–541.
60. Ardila-Hani A, Soffer EE. Review article: the impact of bariatric surgery on gastrointestinal motility. *Aliment Pharmacol Ther*. 2011;34:825–831.
61. Gamagaris Z, Patterson C, Schaye V, et al. Lap-band impact on the function of the esophagus. *Obes Surg*. 2008;18:1268–1272.
62. Mejía-Rivas MA, Herrera-López A, Hernández-Calleros J, et al. Gastroesophageal reflux disease in morbid obesity: the effect of Roux-en-Y gastric bypass. *Obes Surg*. 2008;18:1217–1224.
63. Merrouche M, Sabate JM, Jouet P, et al. Gastro-esophageal reflux and esophageal motility disorders in morbidly obese patients before and after bariatric surgery. *Obes Surg*. 2007;17:894–900.

CHAPTER 10

Postbariatric surgery gastroesophageal reflux disease

Joseph M. Blankush and Joseph R. Broucek
Division of General Surgery, Vanderbilt University Medical Center, Nashville, TN, United States

Introduction

Gastroesophageal reflux disease (GERD) is a common ailment in both the pre- and postbariatric surgery population. Central obesity and elevated BMI have both been shown to independently increase a patient's risk of GERD, erosive esophagitis, and Barrett's esophagus (BE).[1,2] Furthermore, obese patients commonly demonstrate proximal displacement of the squamocolumnar junction and pH transition point within the lower esophageal sphincter (LES) and have a higher frequency of transient LES relaxation compared to patients with normal BMI.[3,4] Various mechanisms have been proposed as explanations for this, including increased intragastric pressure and an increased pressure gradient across the gastroesophageal junction. Obesity has also been associated with increased prevalence of hiatal hernia, impaired gastric emptying, and esophageal dysmotility, all inherently representative of increased risk for reflux disease.[5]

Bariatric surgery, presumably, would then constitute a key intervention for patients with concomitant obesity and GERD. A reduction in central adiposity should conceivably reverse or completely resolve the aforementioned physiologic changes related to obesity that promote reflux disease. While this holds true for a large subset of patients, recent trials have, unfortunately, noted de novo development of GERD in as many as 32% of patients undergoing laparoscopic sleeve gastrectomy (LSG) and 11% of patients undergoing laparoscopic Roux-en-Y gastric bypass (LRYGB).[6] Subsequently, multiple mechanisms related to postbariatric surgery anatomy have been proposed to explain this phenomenon, including decreased gastric volume and compliance creating a high-pressure gastric system, blunting of the angle of His and loss of the angle of His flap valve, and impaired gastric emptying.[7]

Obesity and Esophageal Disorders
DOI: https://doi.org/10.1016/B978-0-323-98365-5.00007-3

While the presence of preoperative erosive esophagitis is associated with a risk of postbariatric surgery GERD, no other clear predisposing patient characteristics have been identified.[8] Conversely, factors related to operative outcomes (e.g., the occurrence of postoperative hiatal hernia) and technique have been identified as clearly predictive of a higher risk of postoperative GERD. In fact, procedure choice and technical aspects of a given bariatric procedure have been found to be the most predictive factor of postoperative GERD. Specifically, LSG has been associated with an odds ratio for the development of GERD as high as 17.9 compared to LRYGB.[9]

So while recent studies have demonstrated similar 5-year weight loss results between LSG and LRYGB, clearly there are differences in outcomes between the two, especially with consideration of both resolution and development of specific morbidities, especially GERD.[6] For this reason, we believe that any consideration of GERD in the postbariatric surgery patient requires discussion of both preoperative work-up and procedure choice in addition to postoperative management.

Pre-operative work up and procedure selection

We believe that establishing both a subjective and objective GERD history is tantamount in providing well-reasoned recommendations to patients regarding the choice of bariatric surgery. This begins with a careful history, identifying signs and symptoms of reflux disease, but this is often easier said than done. Various validated qualitative scoring systems and questionnaires exist to both establish a GERD history as well as quantify the pathologic impact on a patient's quality of life. Example questionnaires include the medical outcomes study 36-item short form (SF-36), the Health-Related Quality of Life Study for GERD (GERD-HRQL), and the QOL questionnaire in reflux and dyspepsia (QOLRAD) to name a few.

The clinical utility of these questionnaires is debatable, though, as studies have noted a discordance between self-reported GERD questionnaire scores and objectives findings on esophagoduodenoscopy (EGD) and/or physiologic studies such as esophageal pH monitoring. As many as 51% of patients who subjectively report severe GERD symptoms do not have objective evidence of pathologic reflux disease.[10]

In practice, patients presenting to our clinic with esophageal symptoms suggestive of GERD such as dyspepsia, globus sensation, dysphagia, and regurgitation are considered likely to have GERD and undergo additional preoperative work-up detailed below.

Beyond the suggestion of reflux symptoms, most society guidelines, including the American Society for Metabolic and Bariatric Surgery (ASMBS) and the International Federation for the Surgery of Obesity (IFSO) recommend EGD before bariatric surgery in symptomatic patients or in those where the surgeon has a suspicion of asymptomatic reflux.[11–13] Notably, several studies have identified decreased esophageal sensitivity in obese patients. An ASMBS metaanalysis demonstrated 6.6%–39.1% (weighted average 16.9%) of preoperative bariatric surgery patients were asymptomatic but had EGD evidence of erosive esophagitis, and 0%–1.3% (0.7%) demonstrated BE.[13] For this reason, we believe that most patients will benefit from EGD evaluation before bariatric surgery, and we liberally apply screening EGD before discussion of an appropriate bariatric procedure for a patient. Specifically, EGD is performed to identify patients with BE, erosive esophagitis, or anatomic considerations such as paraesophageal hernia.

Additional preoperative work-up for any patient about to undergo bariatric surgery is as follows: We recommend that all patients undergo a contrasted, dynamic upper gastrointestinal swallow study (UGI). High-resolution manometry is recommended in patients with clinical symptoms suggestive of dysmotility and/or patients with UGI abnormalities suggestive of primary motility disorders such as achalasia and diffuse esophageal spasm or findings suggestive of secondary motility disorders. Finally, gastric emptying studies are employed in preoperative patients with history and symptoms concerning for delayed gastric emptying, especially in patients with a history of poorly controlled diabetes.

While a lower preoperative DeMeester score may portend a higher likelihood of resolution of GERD following LSG, the pre-LSG DeMeester score has not been shown to accurately predict a patient's likelihood of requiring conversion from LSG to LRYGB.[14,15] Given this, rarely do patients undergo a pH study in our preoperative work-up, but in patients with extraesophageal manifestations such as cough and shortness of breath and with a normal EGD, a pH study may be employed to identify or rule out reflux disease before bypass surgery.

For patients found to have clinical or EGD evidence of GERD during preoperative work-up, we feel it is important to discuss the known data when it comes to resolution, exacerbation, or development of GERD following common bariatric procedures: Most notably, for patients without subjective or objective evidence of GERD, metaanalyses have shown a 23% rate of de novo GERD in LSG and a 2% rate in LRYGB.[16,17] For patients with preoperative GERD symptoms, the SM-BOSS trial demonstrated a worsening of preoperative GERD in 32% of LSG patients and 6% of LRYGB patients.[6]

Ultimately, we typically counsel preoperative bariatric patients as follows: For patients with preexisting, lifestyle-limiting GERD, we overwhelmingly recommend LRYGB unless there are strong indications for LSG. These indications are then weighed, in close consultation with the patient, against the potential for exacerbation of a patient's GERD. We regularly quote data from a recent metaanalysis that demonstrates an average rate of conversion from LSG to LRYGB for intractable postoperative GERD of about 4%.[16] Of note, the Delphi consensus statement on LSG adds that LSG is a suitable option for "patients with severe GERD requiring daily medication if the patient can commit to lifelong endoscopic surveillance."[18] (More on recommendations for endoscopic surveillance below.)

For all patients with erosive esophagitis, LRYGB is recommended unless there is a contraindication or additional reason where sleeve gastrectomy is an optimal procedure. Multiple studies have demonstrated improvement or resolution of esophagitis in most patients following LRYGB while LSG is associated with disease progression.

All patients with long-segment or dysplastic Barrett's esophagus should be considered for preoperative ablative therapy and then recommended for RYGB. Short segment BE overwhelmingly should undergo LRYGB but if significant contraindications to LRYGB exist, LSG with a lifelong commitment to surveillance can be considered.

Postbariatric surgery gastroesophageal reflux disease

Following bariatric surgery, GERD remains a significant lifestyle-limiting morbidity that exists at higher rates for certain procedures, namely, sleeve gastrectomy and duodenal switch procedures, but has been found to be present across all postbariatric surgery patients. A careful consideration and work-up of the potential etiologies specific to each procedure are paramount in choosing the appropriate subsequent intervention.

For all postbariatric patients presenting with subjective complaints of GERD, work-up should begin with a complete history and physical that focuses on pertinent items such as eating habits, persistent morbid obesity, and pattern of weight regain. Simultaneous food and water intake, overeating, and/or eating just before laying down to sleep may significantly contribute to postoperative GERD, as will persistent morbid obesity and persistent increased intraabdominal pressure.

Initial management of reflux following bariatric surgery involves medical therapy, including trials of acid secretion modifying medications such as proton pump inhibitors and H2-receptor antagonists. These are often prescribed in combination with mucosal barriers such as alginate or sucralfate formulations. Very often, medical therapy alleviates reflux symptoms. Even still, patients should be considered for ongoing EGD surveillance.

Subsequent work-up of persistent symptoms, while similar to preoperative work-up, should consider the patient's postbariatric surgery anatomy and the unique etiologies of persistent and de novo GERD attributable to each procedure.

To identify the most likely etiology of a patient's postoperative GERD, we undertake a common pathway for all post-op patients presenting with subjective complaints of reflux symptoms. All patients undergo an upper GI fluoroscopic study (UGI), which we feel gives excellent insight into esophageal dysmotility, anatomic variations contributing to reflux, and in some instances may identify delayed gastric emptying. For patients status-post Roux-en-Y gastric bypass, it is important that the UGI be followed through the jejunojejunostomy (JJ). While bile reflux from the Roux limb is a less likely etiology of reflux in a bariatric surgery patient, a short roux limb or JJ anastomotic stricture may contribute to reflux symptoms.

Following an upper GI series, we obtain EGD to evaluate for both evidence of reflux (i.e. esophagitis, Barrett's esophagus, bile staining, etc.) and further evaluate any suspected contributing anatomic factors (i.e., paraesophageal hernia, anastomotic strictures, and sleeve or pouch abnormalities) as shown in Fig. 10.1.

In select patients with signs of dysphagia and/or suggestion of dysmotility, we will obtain high-resolution manometry to evaluate for esophageal dysmotility that may be contributing to a patient's reflux symptoms. Importantly, in patients without anatomic etiologies of post-LSG GERD, esophageal motility has been demonstrated as the major determinant of sleeve transit rates.[19] While not part of our standard work-up, some practitioners employ Endoflip as a means of simultaneous EGD evaluation and characterization of esophageal motility and lower esophageal sphincter (dys)function.

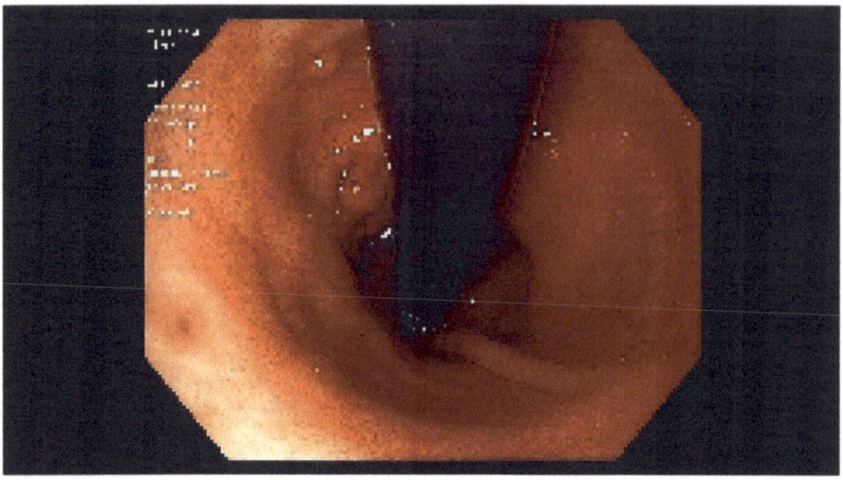

Figure 10.1 Evidence of a hiatal hernia in a patient who had previously undergone laparoscopic sleeve gastrectomy during an upper endoscopy.

Esophageal pH studies are rarely employed in our work-up but do have a place, similar to our preoperative work-up, in patients with a normal EGD and without classic upper GI symptoms of GERD or extraesophageal manifestations. A patient who then goes on to have a normal pH study will often undergo impedance testing to rule out bile reflux as the source of their symptoms.

Finally, vagus nerve injuries are not uncommon in pylorus-preserving bariatric procedures. This can lead to new-onset gastroparesis and delayed gastric emptying that contributes to postbariatric surgery GERD. Patients with UGI and symptoms concerning for delayed gastric emptying will be asked to complete a gastric emptying study as part of their postoperative GERD work-up. Those found to have gastric emptying studies consistent with delayed gastric emptying may initially be referred for serial dilations of the pylorus and/or pyloric sphincter botulinum toxin injections. In many cases, however, reflux symptoms secondary to vagal injury and delayed gastric emptying persist beyond endoscopic interventions and eventually require a conversion to LRYGB.

Postsleeve gastrectomy considerations

Owing to its shorter operative time, lower technical difficulty, and comparable resolution of comorbidities, LSG is now the most commonly performed bariatric procedure in the United States.[20] Surgeons will often cite the preservation of native anatomy and subsequent ability to convert to LRYGB in the future as well as the preserved endoscopic access to the stomach as benefits of LSG over LRYGB. Overwhelmingly, however, the most persistent morbidity following sleeve gastrectomy is reflux disease. As high as 38% of post-LSG patients are found to have esophagitis, and many studies have noted high rates of both symptom exacerbation (19%–37%) as well as de novo GERD (up to 23%) in post-LSG patients.[21] A recent Italian study found a 17% rate of de novo Barrett's and noted an increase in every LA esophagitis grade.[22]

Retained excess weight is of primary consideration in patients with reflux following sleeve gastrectomy. Persistent obesity and intraabdominal pressure undoubtedly contribute to persistent or de novo GERD. In patients with significant persistent central adiposity following medical optimization, we recommend conversion to gastric bypass.

Other anatomic etiologies of persistent or de novo reflux should be considered in postsleeve patients as well. Hiatal hernias, for example, should be closed at the time of the index operation or considered a likely etiology of GERD in the postoperative patient and treated liberally with operative intervention. A large, retained posterior fundus secondary to inadequate gastrectomy at the patient's intiatial procedure can also contibute to post-operative GERD and is also often seen in post-LSG patient with persisent morbid obesity.

Other technical aspects of the sleeve gastrectomy could contribute to GERD, including the presence of a mid-body stenosis (Fig. 10.2), often at the site of the incisura, or a spiraling of the staple line, both common etiologies of increased intragastric pressure. A large, retained posterior fundus secondary to inadequate gastrectomy at the patient's initial procedure can also contibute to post-operative GERD and often seen in post-LSG patient with persisent morbid obesity. In some cases of mid-sleeve stricture, endoscopic pneumatic balloon dilation of the pouch may provide relief without the need for surgical intervention. In many cases, however, patients will require at least a revision of the sleeve if not conversion to LRYGB.

In a subset of postbariatric GERD patients, the etiology of their reflux disease is a persistently hypotensive LES. In these patients, there are a variety of adjunct, minimally invasive interventions that are newer to the market and still largely requiring further study and long-term follow-up but that have been found to provide promising outcomes in carefully selected populations.

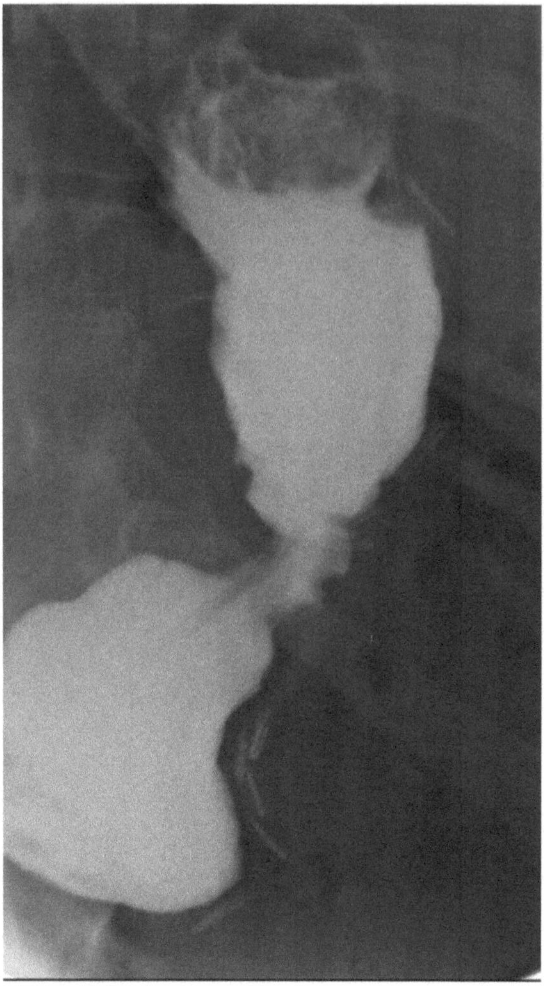

Figure 10.2 Upper GI study demonstrating a mid-gastric sleeve stenosis in a postlaparoscopic sleeve gastrectomy patient.

Magnetic sphincter augmentation, for example, has been proposed as an adjunct procedural intervention for patients with reflux attributable to a hypotensive LES following sleeve gastrectomy. At the time of this writing, magnetic sphincter augmentation devices are not FDA approved for use following gastric sleeve, and only small trials have been undertaken to validate their efficacy in this context. Some small series have noted, however, a resolution of symptoms in a majority of patients who underwent magnetic sphincter augmentation procedures following sleeve gastrectomy.[23–25] The authors would recommend this only in the setting of patients without esophageal dysmotility, proven by a normal high resolution manomoetry study, and who have achieved a normal BMI following sleeve gastrectomy but present with GERD and evidence of a hypotensive LES.

The Stretta device endoscopically provides electrical stimulation in the form of radiofrequency waves to the lower esophageal sphincter, providing a minimally invasive intervention for postbariatric GERD in select patients with a <1% complication rate and requiring no further alteration of anatomy. Several small studies have demonstrated that following the Stretta procedure, the thickness of the esophageal muscle at the gastroesophageal junction (GEJ) increases, the compliance of the GEJ decreases, and there is an associated decrease in the frequency of transient relaxation of the lower esophageal sphincter.[26] Studies to date consist mostly of small case series in the postbariatric surgery population. While results are mixed regarding the long-term efficacy of the Stretta procedure in reducing reflux symptoms, some case series have demonstrated a reduction in GERD symptoms, PPI use, and esophageal acid exposure.[27–29] Similar to patient selection for magnetic sphincter augmentation, other etiologies of GERD such as hiatal hernia and esophageal dysmotility should be ruled out before performance of the procedure.

Finally, antireflux mucosectomy (ARMS) is a more-recent endoscopic intervention that has demonstrated some efficacy in reducing symptoms of reflux disease without the need for further alteration of anatomy or device implantation.[30] In short, a circumferential segment of the gastric cardia is excised and the segment is allowed to scar in, creating a relative stricture at the cardia. This often does require serial balloon dilation of the stricture but in small case series has shown resolution of reflux symptoms. More research and long-term prospective studies, particularly in postbariatric surgery patients, are required before this becomes a mainstay intervention for persistent or de novo reflux.

In short, the only proven, long-term method for the improvement or eradication of GERD following LSG is conversion to LRYGB. Studies with a 2-year follow-up have shown a greater than 90% resolution of GERD symptoms following conversion from LSG to LRYGB.[31] Unfortunately most studies have primarily relied on subjective measures of GERD symptoms and rarely objective EGD evidence of resolution of BE, erosive esophagitis (EE), etc.

Some small series have, however, demonstrated EGD evidence of complete resolution of BE and EE in a majority of patients following conversion to LRYGB. Carandina et al. noted complete resolution of BE and/or EE in 57 of 80 patients (71%).[32] Of the 23 patients without complete resolution following conversion, 19 of them had previously undergone laparoscopic banding. This further demonstrates what multiple studies have shown to date: that laparoscopic banding is associated with high rates of esophageal dilatation and dysmotility disorders.

Postduodenal switch considerations

Foregut anatomy following a duodenal switch procedure is the same as that following a LSG. For this reason, our approach to the initial management, work-up, and subsequent recommendations for intractable GERD are similar to those detailed earlier for post-LSG patients. Conversion to LRYGB is more complex in this patient population given the presence of a GJ anastomosis as well as a second anastomosis between the jejunum and the terminal ileum. For this reason, interventions such as MSA, Stretta, or ARMS that do not require further alteration of the anatomy are more easily employed. In some instances, however, the anatomic variabilities of the gastric sleeve in the duodenal switch may be the primary etiology of postoperative GERD therefore requiring at least a revision of the sleeve if not a conversion to gastric pouch with Roux-en-Y anatomy.

Postgastric band considerations

Overwhelmingly, gastric banding has fallen out of favor as a commonly recommended bariatric procedure. Less than 1% of bariatric procedures performed in 2019 were banding procedures. Still, as recently as 2011, 35% of the procedure performed in the United States were gastric banding, and tens of thousands of patients persist today with an adjustable gastric band in place.[20]

Patients with a laparoscopic band in place who present with GERD often will see an improvement of their symptoms with a deflation of the gastric band, as symptoms are often related to the high-pressure system created in the proximal stomach but the gastric band. Following deflation, especially in patients with persistent obesity, we undertake the aforementioned work-up of patients before any bariatric surgery. Given the significant prevalence of esophageal dysmotility, presumably secondary to persistent esophageal dilation in the setting of gastric banding, we routinely will obtain high-resolution manometry before proceeding with any procedural interventions.

Post-Roux-en-Y gastric bypass considerations

Far fewer patients report severe GERD symptoms following LRYGB. In fact, LRYGB is often a revisional procedure recommended for both bariatric and nonbariatric patients alike who have failed GERD management with previous fundoplication or contraindications to fundoplication.

Still, a systematic review of multiple trials found de novo GERD in 2% of post-LRYGB patients and a worsening of symptoms in 2%–9%. The SM-BOSS trial noted an exacerbation of reflux symptoms in 6% of patients post-LRYGB and de novo GERD in 11%. Notably, however, the same trial found that 60% of post-LRYGB patients experienced a remission of their GERD symptoms.

A properly created gastric pouch, now excluded from the remnant stomach and the pylorus, represents a low-pressure body that should intrinsically enable the forward flow of gastric contents. Furthermore, LRYGB functionally decreases the amount of parietal cells and by extension the acid-producing capabilities of the stomach that remains in continuity with the esophagus. In this context, we assume that reflux is secondary to esophageal dysmotility, a hypotensive LES with distal obstruction, or an another anatomic, mechanical complication of the LRYGB procedure. Some possible structural etiologies of reflux symptoms include the following:

- Hiatal hernia
- Distal obstruction in the form of a gastrojejunostomy or jejunojejunostomy stricture
- A short roux limb causing persistent bile reflux
- A gastro-gastric fistula
- Oversized gastric pouch (normal volume ~15–30 mL).

Initial management focuses on optimizing weight loss and medical management of reflux disease. Subsequent work-up is focused on identifying the aforementioned, reversible abnormalities. EGD is obtained to look for evidence of esophagitis, hiatal hernia, bile reflux, and/or anastomotic stricture. UGI esophagram is also obtained to characterize LES and esophageal (dys)function and further characterize any anatomic abnormalities, including the presence of a gastro-gastric fistula. Although not included in our standard work-up, some practitioners will obtain volumetric CT scans to characterize not only structural abnormalities and fistulas but also to better characterize the measurements and adequacy of the gastric pouch.

Following this work-up and medical and weight optimization, operative intervention may be required. Surgical intervention may be employed to take down a fistula, revise an anastomosis, relieve a distal obstruction, or lengthen the roux limb. If the issue seems to stem from a hypotensive LES, then similar interventions can be employed as when this is found following LSG, including MSA, Stretta, and ARMS, although it is rare for a hypotensive LES to be the sole cause of post-surgical GERD in the absence of a distal narrowing or some other mechanical cause of retrograde flow.

In our practice, we do not regularly perform the single anastomosis gastric bypass, a procedure in which a gastric pouch is made and then a loop of jejunum is brought up to the pouch and a single GJ anastomosis is performed. Bile reflux is a common morbidity following this procedure and should be considered a primary cause of reflux symptoms in patients who have undergone this procedure, and EGD will often demonstrate bile reflux in this population. Conversion to Roux-en-Y anatomy may be required for the resolution of reflux symptoms.

Postoperative surveillance

Studies have demonstrated that postsleeve gastrectomy patients with de novo development of BE do not reliably experience symptoms of reflux. Furthermore, studies investigating the long-term rates of BE in LSG patients range from 4% to 17%, which far exceeds the incidence of BE noted on preoperative EGD (~2%).[16,33] Therefore we recommend that post-LSG patients undergo surveillance EGDs, regardless of reflux symptoms, at 1, 3, and 5 years post-op and every 3–5 years after depending on visual examination and biopsy results.

Conclusion

GERD remains an important morbidity in both the pre- and postbariatric surgery population. While bariatric surgery is an effective tool for the treatment of GERD in the obese population, careful preoperative GERD work-up and patient counseling are required in determining the appropriate procedure for each patient. Postbariatric surgery GERD is often secondary to anatomic changes related to either the procedure itself

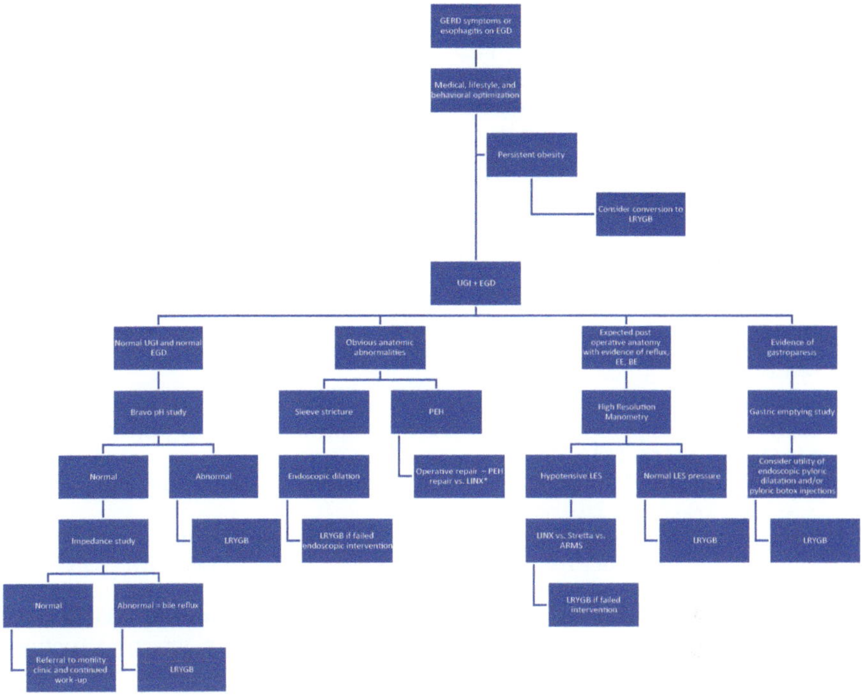

Figure 10.3 Proposed algorithm for management of GERD in bariatric population. *GERD*, Gastroesophageal reflux disease; *LRYGB*, laparoscopic Roux-en-Y gastric bypass; *UGI*, fluoroscopic contrasted upper gastrointestinal study; *EGD*, esophagoduodenoscopy; *EE*, erosive esophagitis; *BE*, Barrett's esophagus; *PEH*, paraesophageal hernia; *LES*, lower esophageal sphincter. *All patients require high-resolution manometry before undergoing the LINX procedure.

(e.g., gastric banding, LSG) or technical complications unique to the individual procedures and should be evaluated accordingly. Fig. 10.3 shows the proposed algorithm for management in this group of patients.

Overall, patients with evidence of preoperative, life-limiting GERD should likely undergo LRYGB. While LSG remains an effective and widespread intervention for obesity and has shown a decrease in postoperative GERD in some patients, the high incidence of worsening or de novo GERD in the post-LSG populations suggests that this population should be surveilled carefully in the postoperative setting. Ultimately, postbariatric surgery patients with intractable GERD have multiple treatment options available to them. To date, and in consideration of the available long-term data, conversion to Roux-en-Y anatomy remains the most advised option.

References

1. El-Serag HB, Hashmi A, Garcia J, et al. Visceral abdominal obesity measured by CT scan is associated with an increased risk of Barrett's oesophagus: a case-control study. *Gut*. 2014;63(220):2–9.
2. Hampel H, Abraham NS, El-Serag HB. Meta-analysis: obesity and the risk for gastroesophageal reflux disease and its complications. *Ann Intern Med*. 2005;143:199–211.
3. Merrouche M, Sabaté J-M, Jouet P, et al. Gastro-esophageal reflux and esophageal motility disorders in morbidly obese patients before and after bariatric surgery. *Obes Surg*. 2007;17:894–900.
4. Robertson EV, Derakhshan MH, Wirz AA, et al. Central obesity in asymptomatic volunteers is associated with increased intrasphincteric acid reflux and lengthening of the cardiac mucosa. *Gastroenterology*. 2013;145:730–739.
5. Chang P, Friedenberg F. Obesity and GERD. *Gastroenterol Clin North Am*. 2014;43:161–173.
6. Peterli R, Wölnerhanssen BK, Peters T, et al. Effect of laparoscopic sleeve gastrectomy vs laparoscopic Roux-en-Y gastric bypass on weight loss in patients with morbid obesity. *JAMA*. 2018;319:255.
7. Ashrafi D, Osland E, Memon MA. Bariatric surgery and gastroesophageal reflux disease. *Ann Transl Med*. 2020;8:S11. -S11.
8. Althuwaini S, Bamehriz F, Aldohayan A, et al. Prevalence and predictors of gastroesophageal reflux disease after laparoscopic sleeve gastrectomy. *Obes Surg*. 2018;28:916–922.
9. Navarini D, Madalosso CAS, Tognon AP, Fornari F, Barão FR, Gurski RR. Predictive factors of gastroesophageal reflux disease in bariatric surgery: a controlled trial comparing sleeve gastrectomy with gastric bypass. *Obes Surg*. 2020;30: 1360–1367.
10. Chan K, Liu G, Miller L, et al. Lack of correlation between a self-administered subjective GERD questionnaire and pathologic GERD diagnosed by 24-h esophageal pH monitoring. *J Gastrointest Surg*. 2010;14:427–436.
11. Di Lorenzo N, Antoniou SA, Batterham RL, et al. Clinical practice guidelines of the European association for endoscopic surgery (EAES) on bariatric surgery: update 2020 endorsed by IFSO-EC, EASO and ESPCOP. *Surg Endosc*. 2020;34:2332–2358.
12. Mechanick JI, Apovian C, Brethauer S, et al. Clinical practice guidelines for the perioperative nutrition, metabolic, and nonsurgical support of patients undergoing bariatric procedures – 2019 update: cosponsored by American association of clinical endocrinologists/American college of endocrinology. *Obesity*. 2020;28:O1–O58.
13. Campos GM, Mazzini GS, Altieri MS, et al. ASMBS position statement on the rationale for performance of upper gastrointestinal endoscopy before and after metabolic and bariatric surgery. *Surg Obes Relat Dis*. 2021;17:837–847.
14. Summe KL, Hawasli A. Can lower preoperative 48-hour pH score predict reflux resolution after sleeve gastrectomy. *Am J Surg*. 2021;221:578–580.
15. De Montrichard M, Greilsamer T, Jacobi D, Bruley Des Varannes S, Mirallié E, Blanchard C. Predictive value of preoperative DeMeester score on conversion to Roux-en-Y gastric bypass for gastroeosophageal reflux disease after sleeve gastrectomy. *Surg Obes Relat Dis*. 2020;16:1219–1224.
16. Yeung KTD, Penney N, Ashrafian L, Darzi A, Ashrafian H. Does sleeve gastrectomy expose the distal esophagus to severe reflux?: a systematic review and meta-analysis. *Ann Surg*. 2020;271:257–265.
17. Gu L, Chen B, Du N, et al. Relationship between bariatric surgery and gastroesophageal reflux disease: a systematic review and meta-analysis. *Obes Surg*. 2019;29: 4105–4113.

18. Mahawar KK, Omar I, Singhal R, et al. The first modified Delphi consensus statement on sleeve gastrectomy. *Surg Endosc*. 2021;35:7027–7033.
19. Johari Y, Wickremasinghe A, Kiswandono P, et al. Mechanisms of esophageal and gastric transit following sleeve gastrectomy. *Obes Surg*. 2021;31:725–737.
20. Alalwan AA, Friedman J, Park H, Segal R, Brumback BA, Hartzema AG. US national trends in bariatric surgery: a decade of study. *Surgery*. 2021;170:13–17.
21. Matar R, Maselli D, Vargas E, et al. Esophagitis after bariatric surgery: large cross-sectional assessment of an endoscopic database. *Obes Surg*. 2020;30:161–168.
22. Genco A, Soricelli E, Casella G, et al. Gastroesophageal reflux disease and Barrett's esophagus after laparoscopic sleeve gastrectomy: a possible, underestimated long-term complication. *Surg Obes Relat Dis*. 2017;13:568–574.
23. Leeds SG, Ngov A, Gerald, Ward MA. Safety of magnetic sphincter augmentation in patients with prior bariatric and anti-reflux surgery. *Surg Endosc*. 2021;35:5322–5327.
24. Kuckelman JP, Phillips CJ, Derickson MJ, Faler BJ, Martin MJ. Esophageal magnetic sphincter augmentation as a novel approach to post-bariatric surgery gastroesophageal reflux disease. *Obes Surg*. 2018;28:3080–3086.
25. Broderick RC, Smith CD, Cheverie JN, et al. Magnetic sphincter augmentation: a viable rescue therapy for symptomatic reflux following bariatric surgery. *Surg Endosc*. 2020;34:3211–3215.
26. Arts J, Bisschops R, Blondeau K, et al. A double-blind sham-controlled study of the effect of radiofrequency energy on symptoms and distensibility of the gastro-esophageal junction in GERD. *Am J Gastroenterol*. 2012;107:222–230.
27. Khidir N, Angrisani L, Al-Qahtani J, Abayazeed S, Bashah M. Initial experience of endoscopic radiofrequency waves delivery to the lower esophageal sphincter (Stretta Procedure) on symptomatic gastroesophageal reflux disease post-sleeve gastrectomy. *Obes Surg*. 2018;28:3125–3130.
28. Mattar SG, Qureshi F, Taylor D, Schauer PR. Treatment of refractory gastroesophageal reflux disease with radiofrequency energy (Stretta) in patients after Roux-en-Y gastric bypass. *Surg Endosc*. 2006;20:850–854.
29. Borbely Y, Bouvy N, Schulz HG, Rodriguez LA, Ortiz C, Nieponice A. Electrical stimulation of the lower esophageal sphincter to address gastroesophageal reflux disease after sleeve gastrectomy. *Surg Obes Relat Dis*. 2018;14:611–615.
30. Inoue H, Ito H, Ikeda H, et al. Anti-reflux mucosectomy for gastroesophageal reflux disease in the absence of hiatus hernia: a pilot study. *Ann Gastroenterol*. 2014;27:346–351.
31. Matar R, Monzer N, Jaruvongvanich V, et al. Indications and outcomes of conversion of sleeve gastrectomy to Roux-en-Y gastric bypass: a systematic review and a meta-analysis. *Obes Surg*. 2021;30:3936–3946.
32. Carandina S, Soprani A, Montana L, et al. Conversion of sleeve gastrectomy to Roux-en-Y gastric bypass in patients with gastroesophageal reflux disease: results of a multicenter study. *Surg Obes Relat Dis*. 2020;16:732–737.
33. Kassir R, Kassir R, Deparseval B, et al. Routine surveillance endoscopy before and after sleeve gastrectomy? *World J Gastrointest Endosc*. 2019;11:1–4.

Index

Note: Page numbers followed by "*f*" and "*t*" refer to figures and tables, respectively.

FSC
www.fsc.org
MIX
Papier | Fördert gute Waldnutzung
FSC® C083411

Kolibri 360 GmbH
Ferdinand-Jühlke-Straße 7
99095 Erfurt, Deutschland
produktsicherheit@kolibri360.de